The Best American Science and Nature Writing 2001

To Peter & Karen
From Christine

The Best American Science and Nature Writing 2001

Edited and with an Introduction
by Edward O. Wilson

Burkhard Bilger, Series Editor

HOUGHTON MIFFLIN COMPANY
BOSTON · NEW YORK 2001

ISSN 1530-1508
ISBN 0-618-08296-4
ISBN 0-618-15359-4 (pbk.)

Printed in the United States of America

DOC 10 9 8 7 6 5 4 3 2 1

Contents

Foreword

THE MAGAZINE RACK in any large American bookstore is one of the wonders of the age: its titles multitudinous, its covers seductive, its contents perpetually renewed, revised, edited, fact-checked, and copyedited. The best work here is as eloquent as any fiction, a step ahead of the nightly news, and more authoritative than any book. It distills an overwhelming tide of information, refracts it through some of the country's best minds, and then does it all over again — daily, weekly, monthly, and bimonthly. It's a marketplace of words, where wit and insight are the gold standard, if not always the currency, and the Zeitgeist alone determines the interest rate.

Still, there were times this year when I could hardly face all those toothy smiles and urgent headlines again. The key phrase above is "best work" — and how small a percentage that really represents. For every *Harper's* or *Atlantic Monthly* there are three dog-grooming magazines out there. For every *Scientific American* there are a dozen devoted to UFOs and the healing power of magnets. The true wonder of the newsstand, a cynic could say, is of so much effort lavished on such trivialities; so many unhealthy appetites catered to so assiduously; so many hours, forests, dollars, and minds wasted so casually.

Both the misery and the majesty of modern publishing, it seems to me, are equal justifications for anthologies like this one. Simply put, there is more good writing out there than ever before, and it is harder and harder to find it. And if that's true for journalism of all kinds, it's doubly true for science and nature writing. The premier journals and magazines in the field — *Nature, Science, Cell,* and so forth — have never been great sources for literary journalism, nor are they meant to be. Instead, the best essays and articles are scattered among a bewildering array of literary and general interest magazines. Edward Wilson and I had a few strict criteria in assembling this volume: no fiction, poetry, prose poems, book chapters (unless published as stand-alone articles), or plays; only nonfiction published in the last calendar year. But that still left thousands of issues and articles to sift and sort.

I've heard tales of other editors in the Best American series renovating whole basements to accommodate their hundreds of subscriptions, building mailboxes as big as sea chests, and tipping their postal workers extravagantly. Living in Brooklyn in a typically cramped brownstone, I don't have the room for that sort of thing, so my method tends to be a bit more concentrated (read frantic). For three months this past winter, I spent most of my days in the library, harassing the periodical room people, leafing through teetering stacks of magazines, photocopying madly as I went. When I was done, I sent some 120 articles to the guest editor, who chopped that number down to 25 or so. We then went over his choices, culling one or two and restoring others, and ended up with the lineup you now see.

As you might expect — or even hope — from a book guest edited by Edward Wilson, this year's selections lean somewhat toward the natural sciences. Geology, physics, mathematics, and chemistry are all represented, but they are surrounded on all sides by crocodiles, harpy eagles, great apes, and a host of other creatures micro- and macroscopic. The result, I think, is a vindication of an oft-maligned field and a hopeful glimpse of its future. Only a year ago, *High Country News* — one of the clearest, least sentimental voices in the environmental field — published an essay entitled "Enough Nature Writing Already!" The author, Stephen Lyons, admitted to being a nature writer himself, but he was willing to sacrifice himself for the greater good. "This may be heresy," he wrote,

> but how many times do we need to wade through an introvert's musings on his or her latest tramp into unspoiled wilderness? Would it hurt anyone to have a moratorium on the word "sacred," or on the following:

> "I take a step slowly across the knoll. I listen to coyotes howl. I watch hawks circle on thermals that I feel against my skin, which is attached to my body. If only all of humankind could walk with me and think the same thoughts I have, then all conflicts, cruelty, and madness would cease."

I know what he means. Our literary magazines sometimes seem like asylums of pain, full of grieving children, recovering alcoholics, and victims of abuse, most of whom seem to seek solace in nature. Their language — burnished to an unearthly glow by years of creative writing courses and readings in Raymond Carver and Barry Lopez — too often serves no larger idea or public purpose. And yet some of the most moving pieces I've read this year are of exactly this sensitive, inward-looking variety (Ted Kerasote's "A Killing at Dawn," for instance). And others, like Mark Cherrington's "To Save a Watering Hole," Edwin Dobb's "New Life in a Death Trap," and David Quammen's "Megatransect," hardly have time for personal epiphanies. Nature, in these essays, is a battlefield, a laboratory, a constantly receding frontier — anything but a refuge.

Lyons concluded his essay by suggesting a ten-year sabbatical, during which "the writer would have to participate in cutting down the exact number of trees responsible to produce his or her work." But this book hints at a more fruitful alternative: have them write about science. A little time spent searching for extremophile bacteria in Antarctica (as Oliver Morton did in "Ice Station Vostok") could go a long way toward airing out those soggy sensibilities. And science journalism, with all its certainties and declarative sentences, could use a few writers with a gift for introspection. The two fields are just across the fence from each other, after all; a little cross-pollinating might improve the harvest all around.

Once again, this year, I've been exceedingly lucky in my guest editor. I've often heard Edward Wilson described as a southern gentleman — as courtly in his manners as he is elegant and incisive in his writing — but that hardly conveys his real warmth and modesty or his ease and thoughtfulness as a collaborator. This volume clearly bears the imprint of his mind, and it's all the better for it.

I'd also like to thank Laura Van Dam, my editor at Houghton Mifflin, for her sharp eye, kind heart, and unflagging good cheer, and Deanne Urmy for shepherding the book in its final stages. Most of all, my love and thanks go to my wife, Jennifer, who had to put up with all those photocopies; to my son, Hans, who has taken to adding blurbs and author notes to his first-grade assignments; and to my daughter, Ruby, who would have liked to have seen more pieces about bones, and bears, and unicorns in this volume. Maybe next year.

Future submissions should be sent, with a very brief cover letter, to Tim Folger, c/o Editor, The Best American Science and Nature Writing 2002, Houghton Mifflin Company, 222 Berkeley Street, Boston, MA 02116.

Burkhard Bilger

Introduction: Life Is a Narrative

LET ME TELL you a story. It is about two ants. In the early 1960s, when I was a young professor of zoology at Harvard University, one of the vexing mysteries of evolution was the origin of ants. That was far from a trivial problem in science. Ants are the most abundant of insects, the most effective predators of other insects, and the busiest scavengers of small dead animals. They transport the seeds of thousands of plant species, and they turn and enrich more soil than earthworms. In totality (they number roughly in the million billions and weigh about as much as all of humanity), they are among the key players of Earth's terrestrial environment. Of equal general interest, they have attained their dominion by means of the most advanced social organization known among animals.

I had chosen these insects for the focus of my research. It was the culmination of a fascination that dated back to childhood. Now, I spent a lot of time thinking about how they came to be. At first the problem seemed insoluble, because the oldest known ants, found in fossil deposits up to 57 million years old, were already advanced anatomically. In fact, they were quite similar to the modern forms all about us. And just as today, these ancient ants were among the most diverse and abundant of insects. It was as though an opaque curtain had been lowered to block our view of everything that occurred before. All we had to work with was the tail end of evolution.

Somewhere in the world the Ur-ants awaited discovery. I had many conversations with William L. Brown, a friend and fellow myrmecologist, about where the missing links might turn up and what traits they possess that could reveal their ancestry among the nonsocial wasps. We guessed that they first appeared in the late Mesozoic era, 65 million or more years ago, far back enough to have stung and otherwise annoyed the last of the dinosaurs. We were not willing to accept the alternative hypothesis favored by some biblical creationists, that ants did not evolve at all but appeared on Earth full-blown.

Because well-preserved fossils had already been collected by the tens of thousands from all around the northern hemisphere over a period of two centuries without any trace of the Ur-species, I was afraid I would never see one in my lifetime. Then, as so often happens in science, a chance event changed everything. One Sunday morning in 1967, a middle-aged couple, Mr. and Mrs. Edmund Frey, were strolling along the base of the seaside bluffs at Cliffwood Beach, New Jersey, collecting bits of fossilized wood and amber from a thin layer of clay freshly exposed by a storm the day before. They were especially interested in the amber, which are jewel-like fragments of fossil tree sap. In one lump they rescued, clear as yellow glass, were two beautifully preserved ants. At first, that might have seemed nothing unusual: museums, including the one at Harvard, are awash in amber ants. What made these specimens important, however, was their age: about 90 million years, from the middle of the Cretaceous period, Mesozoic era, in the Age of Dinosaurs.

The Freys were willing to share their find, and soon the two specimens found their way to me for examination. There they came close to disaster. As I nervously fumbled the amber piece out of its mailing box I dropped it to the floor, where it broke into two halves. Luck stayed with me, however. The break was as clean as though made by a jeweler, and each piece contained an undamaged specimen. Within minutes I determined that the ants were the long-sought Holy Grail of ant paleontology, or at least very close to it. Brown and I later formally placed them in a new genus, *Sphecomyrma freyi* (literally, "Frey's wasp ant"). They were more primitive than all other known ants, living and fossil. Moreover, in a dramatic confirmation of evolution as a predictive theory, they possessed most of the intermediate traits that according to our earlier deductions should connect modern ants to the nonsocial wasps.

As a result of the discovery, other entomologists intensified their search, and many more ant fossils of Mesozoic age were soon found. Originating from deposits in New Jersey, Canada, Siberia, and Brazil, they compose a mix of primitive and more advanced species. Bit by bit, they have illuminated the history of ants from near the point of origin over 100 million years ago to the start of the great radiative spread that created the modern fauna.

Science consists of millions of stories like the finding of New Jersey's dawn ants. These accounts, some electrifying, most pedestrian, become science when they can be tested and woven into cause-and-effect explanations to become part of humanity's material worldview. Science, like the rest of culture, is based on the manufacture of narrative. That is entirely natural, and in a profound sense it is a Darwinian necessity. We all live by narrative, every day and every minute of our lives. Narrative is the human way of working through a chaotic and unforgiving world bent on reducing our bodies to malodorous catabolic molecules. It delays the surrender of our personal atoms and compounds back to the environment for the assembly of more humans, and ants.

By narrative we take the best stock we can of the world and our predicament in it. What we see and recreate is seldom the blinding literal truth. Instead, we perceive and respond to our surroundings in narrow ways that most benefit our organismic selves. The narrative genius of *Homo sapiens* is an accommodation to the inherent inability of the three pounds of our sensory system and brain to process more than a minute fraction of the information the environment pours into them. In order to keep the organism alive, that fraction must be intensely and accurately selective. The stories we tell ourselves and others are our survival manuals.

With new tools and models, neuroscientists are drawing close to an understanding of the conscious mind as narrative generator. They view it as an adaptive flood of scenarios created continuously by the working brain. Whether set in the past, present, or future, whether fictive or reality based, the free-running constructions are our only simulacrum of the world outside the brain. They are everything we will ever possess as individuals. And, minute by minute, they determine whether we live or die.

The present in particular is constructed from sensations very far in excess of what can be put into the simulacrum. Working at a frantic pace, the brain summons memories — past scenarios — to help screen and organize the incoming chaos. It simultaneously creates imaginary scenarios to create fields of competing options, the process we call decision-making. Only a tiny fraction of the narrative fragments — the focus — is selected for higher-order processing in the prefrontal cortex. That segment constitutes the theater of running symbolic imagery we call the conscious mind.

During the story-building process, the past is reworked and returned to memory storage. Through repeated cycles of recall and supplementation the brain holds on to shrinking segments of the former conscious states. Across generations the most important among these fragments are communicated widely and converted into history, literature, and oral tradition. If altered enough, they become legend and myth. The rest disappear. The story I have just told you about Mesozoic ants is all true as best I can reconstruct it from my memory and notes. But it is only a little bit of the whole truth, most of which is beyond my retrieval no matter how hard I might try.

This brings me to the relation between science and literature. Science is not a subculture separate from that of literature. Its knowledge is the totality of what humanity can verify about the real world, testable by repeated experiment or factual observation, bound to related information by general principles, and — this is the part most often missed — ultimately subject to cause-and-effect explanations consilient across the full range of disciplines. The most democratic of human mental activity, it comprises the nonfiction stories you can take to the bank.

Everyone can understand the process of science, and, once familiar with a modest amount of factual information and the elementary terminology of particular disciplines, he or she can grasp the intuitive essence of at least some scientific knowledge. But the scientific method is not natural to the human mind. The phenomena it explicates are by and large unfamiliar to ordinary experience. New scientific facts and workable theories, the silver and gold of the scientific enterprise, come slow and hard, less like nuggets lying on a streambed than ore dug from mines. To enjoy them while maintaining an effective critical attitude requires mental discipline. The reason, again, is the innate constraints of the human brain. Gossip and music flow easily through the human mind, because the brain is genetically predisposed to receive them. Theirs is a Paleolithic cogency. Calculus and reagent chemistry, in contrast, come hard, like ballet on pointe. They have become relevant only in modern, postevolutionary times. Of the hundreds of fellow scientists I have known for more than fifty years, from graduate students to Nobelists, all generally prefer at random moments of their lives to listen to gossip and music rather than to scientific lectures. Trust me: physics is hard even for physicists. Somewhere on a distant planet, there may exist a species that hereditarily despises gossip and thrives on calculus. But I doubt it.

The central task of science writing for a broad audience is, in consequence, how to make science human and enjoyable without betraying nature. The best writers achieve that end by two means. They present the phenomena as a narrative, whether historical, evolutionary, or phenomenological, and they treat the scientists as protagonists in a story that contains, at least in muted form, the mythic elements of challenge and triumph.

To wring honest journalism and literature from honest science, the writer must overcome formidable difficulties. First is the immensity and exponential growth of the primary material itself, which has experienced a phenomenally short doubling time of fifteen years for over three hundred years, all the while coupled with a similarly advancing technology. It has spread its reach into every conceivable aspect of material existence, from the origin of the universe to the creative process of the mind itself. Its relentless pursuit of detail and theory long ago outstripped the minds of individual scientists themselves to hold it. So fragmented are the disciplines and specialized the language resulting from the growth that experts in one subject often cannot grasp the technical reports of experts in closely similar specialties. Insect neuroendocrinologists, for example, have a hard time understanding mammalian neuroendocrinologists, and the reverse. To see this change in science graphically you need only place opened issues of a premier journal such as *Nature* or *Science* from fifty years ago side by side with issues of the same journal today. The science writer must somehow thread his way into this polyglot activity, move to a promising sector of the front, and, then, accepting a responsibility the research scientists themselves typically avoid, turn the truth of it into a story interesting to a broad public.

A second obstacle to converting science into literature is the standard format of research reportage in the technical journals. Scientific results are by necessity couched in specialized language, trimmed for brevity, and delivered raw. Metaphor is unwelcome except in small homeopathic doses. Hyperbole, no matter how brilliant, spells death to a scientific reputation. Understatement and modesty, even false modesty, are preferred, because in science discovery counts for everything and personal style next to nothing.

In pure literature, metaphor and personal style are, in polar contrast, everything. The creative writer, unlike the scientist, seeks channels of cognitional and emotional expression already deeply carved by instinct and culture. The most successful innovator in literature is an honest illusionist. His product, as Picasso said of visual art, is the lie that helps us to see the truth. Imagery, phrasing, and analogy in pure literature are not crafted to report empirical facts. They are instead the vehicles by which the writer transfers his own feelings directly into the minds of his readers in order to evoke the same emotional response.

The central role of pure literature is the transmission of the details of human experience by artifice that directs aesthetic reaction. Originality and power of metaphor, not new facts and theory, are coin of the realm in creative writing. Their source is an intuitive understanding of human nature as opposed to an accurate knowledge of the material world, at least in the literal, quantifiable form required for science. Metaphor in the best writing strikes the mind in an idiosyncratic manner. Its effect ripples out in a hypertext of culture-bound meaning, yet it triggers emotions that transcend culture. Technical scientific reporting tries to achieve exactly the reverse: it narrows meaning and avoids metaphor in order to preserve literalness and repeatability. It saves emotional resonance for another day and venue.

To illustrate the difference, I've contrived the following imaginary examples of the two forms of writing applied to the same subject, the search for life in a deep cave:

> SCIENCE. The central shaft of the cavern descends from the vegetated rim to the oblique slope of fallen rock at the bottom, reaching a maximum depth of 86 meters before giving way to a lateral channel. On the floor of this latter passageway we found a small assemblage of troglobitic invertebrates, including two previously undescribed eyeless species of the carabid subfamily Bembidini (see also Harrison, in press).

> LETTERS. After an hour's rappel through the Hadean darkness we at last reached the floor of the shaft almost 300 feet below the fern-lined

> rim. From there we worked our way downward across a screelike rubble to the very bottom. Our headlamps picked out the lateral cavern exactly where Romero's 1926 map claimed it to be. Rick pushed ahead and within minutes shouted back that he had found blind, white cave inhabitants. When we caught up, he pointed to scurrying insects he said were springtails and, to round out the day, at least two species of ground beetles new to science.

In drawing these distinctions in the rules of play, I do not mean to depict scientists as stony Pecksniffs. Quite the contrary. They vary enormously in temperament, probably to the same degree as a random sample from the nonscientific population. Their conferences and seminars are indistinguishable in hubbub from business conventions. Nothing so resembles an ecstatic prospector as a scientist with an important discovery to report to colleagues, to family, to grant officers, to anyone who will listen.

A scientist who has made an important discovery is as much inclined to show off and celebrate as anyone else. Actually, this can be accomplished in a technical article, if done cautiously. The heart of such a report is always the Methods and Materials, followed by Results, all of which must read like your annual tax report. But up front there is also the Introduction, where the author briefly explains the significance of the topic, what was known about it previously, who made the previous principal advances, and what aspect of the whole the author's own findings are meant to address. A smidgen of excitement, maybe even a chaste metaphor or two, is allowed in the Introduction. Still more latitude is permitted for the Discussion, which follows the Results. Here the writer is expected to expatiate on the data and hypotheses as inclination demands. He or she may also push the envelope and make cautious guesses about what lies ahead for future researchers. However, there must be no outbursts such as, "I was excited to find . . ." or "This is certain to be a major advance."

Science writers are in the difficult position of locating themselves somewhere between the two stylistic poles of literature and science. They risk appearing both as journalists to the literati and as amateurs to the scientists. But these judgments, if made, are ignorant and unfair. Enormous room for original thought and expression exists in science writing. Its potential is nothing less than the establishment of what Sir Charles Snow called the third culture, a concept also recently promoted by the author and literary agent John Brockman.

The position nearest the literary pole is that broadly classified as nature writing. With roots going back to nineteenth-century romanticism, it cultivates the facts and theories of science but relies heavily on personal narrative and aesthetic expression. Thanks to writers of the first rank such as Annie Dillard, Barry Lopez, Peter Matthiessen, Bill McKibben, David Quammen, and Jonathan Weiner (a representative but far from exclusive list!), nature writing has become a distinctive American art form.

The pole nearest science is occupied primarily by scientists who choose to deliver their dispatches from the front to a broader public. Ranging from memoirs to philosophical accounts of entire disciplines, their writing resonates with a certain firsthand authority but is constrained to modesty in emotional expression by the conventions of their principal trade. Writing scientists also frequently struggle with the handicap imposed by the lack of connection of their subject to ordinary human experience: few tingles of the spine come from bacterial genetics, and generally the only tears over physical chemistry come as a result of trying to learn it.

Despite the inherent difficulties, science writing is bound to grow in influence, because it is the best way to bridge the two cultures into which civilization is still split. Most educated people who are not professionals in the field do not understand science and technology, despite the profound effect of these juggernauts of modernity on every aspect of their lives. Symmetrically, most scientists are semiliterate journeymen with respect to the humanities. They are thus correspondingly removed from the heart and spirit of our species. How to solve this problem is more than just a puzzle for creative writers. It is, if you will permit a scientist a strong narrative-laden metaphor, the central challenge of education in the twenty-first century.

EDWARD O. WILSON

The Best American Science and Nature Writing 2001

DAVID BERLINSKI

Iterations of Immortality

FROM *Harper's Magazine*

THE CALCULUS and the rich body of mathematical analysis to which it gave rise made modern science possible, but it was the algorithm that made possible the modern world. They are utterly different, these ideas. The calculus serves the imperial vision of mathematical physics. It is a vision in which the real elements of the world are revealed to be its elementary constituents: particles, forces, fields, or even a strange fused combination of space and time. Written in the language of mathematics, a single set of fearfully compressed laws describes their secret nature. The universe that emerges from this description is alien, indifferent to human desires.

The great era of mathematical physics is now over; the 300-year effort to represent the material world in mathematical terms has exhausted itself. The understanding that it was to provide is infinitely closer than it was when Isaac Newton wrote in the late seventeenth century, but it is still infinitely far away. One man ages as another is born, and if time drives one idea from the field, it does so by welcoming another. The algorithm has come to occupy a central place in our imagination. It is the second great scientific idea of the West. There is no third.

An algorithm is an *effective* procedure — a recipe, a computer program — a way of getting something done in a finite number of discrete steps. Classical mathematics contains algorithms for virtually every elementary operation. Over the course of centuries, the complex (and counterintuitive) operations of addition, multiplication, subtraction, and division have been subordinated to fixed routines. Arithmetic algorithms now exist in mechanical form; what was once an intellectual artifice has become an instrumental artifact.

The world the algorithm makes possible is retrograde in its nature to the world of mathematical physics. Its fundamental theoretical objects are symbols, and not muons, gluons, quarks, or space and time fused into a pliant knot. Algorithms are human artifacts. They belong to the world of memory and meaning, desire and design. The idea of an algorithm is as old as the dry, humped hills, but it is also cunning, disguising itself in a thousand protean forms. It was only in this century that the concept of an algorithm was coaxed completely into consciousness. The work was undertaken more than sixty years ago by a quartet of brilliant mathematical logicians: Kurt Gödel, Alonzo Church, Emil Post, and A. M. Turing, whose lost eyes seem to roam anxiously over the second half of the twentieth century.

If it is beauty that governs the mathematician's soul, it is truth and certainty that remind him of his duty. At the end of the nineteenth century, mathematicians anxious about the foundations of their subject asked themselves why mathematics was true and whether it was certain, and to their alarm discovered that they could not say and did not know. Caught between mathematical crises and their various correctives, logicians were forced to organize a new world to rival the abstract, cunning, and continuous world of the physical sciences, their work transforming the familiar and intuitive but hopelessly unclear concept of the algorithm into one both formal and precise.

Unlike Andrew Wiles, who spent years searching for a proof of Fermat's last theorem, the logicians did not set out to find the concept that they found. They were simply sensitive enough to see what they spotted. We still do not know why mathematics is true and whether it is certain. But we know what we do not know in an immeasurably richer way than we did. And learning this has been a remarkable achievement, among the greatest and least known of the modern era.

Dawn kisses the continents one after the other, and as it does a series of coded communications hustles itself along the surface of the earth, relayed from point to point by fiber-optic cables, or bouncing in a triangle from the earth to synchronous satellites, serene in the cloudless sky, and back to earth again, the great global network of computers moving chunks of data at the speed of light: stock-market indexes, currency prices, gold and silver futures, news of cotton crops, rumors of war, strange tales of sexual scandal, images of men in starched white shirts stabbing at keyboards with stubby fingers or looking upward at luminescent monitors, beads of perspiration on their tensed lips. E-mail flashes from server to server, the circle of affection or adultery closing in an electronic braid; there is good news in Lisbon and bad news in Saigon. There is data everywhere and information on every conceivable topic: the way raisins are made in the Sudan, the history of the late Sung dynasty, telephone numbers of dominatrixes in Los Angeles, and pictures too. A man may be whipped, scourged, and scoured without ever leaving cyberspace; he may satisfy his curiosity or his appetites, read widely in French literature, decline verbs in Sanskrit, or scan an interlinear translation of the *Iliad,* discovering the Greek for *greave* or *grieve;* he may search out remedies for obscure diseases, make contact with covens in South Carolina, or exchange messages with people in chat groups who believe that Princess Diana was murdered on instructions tendered by the House of Windsor, the dark demented devious old Queen herself sending the order that sealed her fate.

All of this is very interesting and very new — indeed, interesting because new — but however much we may feel that our senses are brimming with the debris of data, the causal nexus that has made the modern world extends in a simple line from the idea of an algorithm, as logicians conceived it in the 1930s, directly to the ever present, always moving now; and not since the framers of the American Constitution took seriously the idea that all men are created equal has an idea so transformed the material conditions of life, the expectations of the race. It is the algorithm that rules the world itself, insinuating itself into every device and every discussion or diagnosis, offering advice and making decisions, maintaining its presence in every transaction, carrying out dizzying computations, arming and then aiming cruise missiles, bringing the dinosaurs back to life on film, and, like blind Tiresias, foretelling the extinction of the universe either in a cosmic crunch or in one of those flaccid affairs in which after a long time things just peter out.

*

The algorithm has made the fantastic and artificial world that many of us now inhabit. It also seems to have made much of the natural world, at least that part of it that is alive. The fundamental act of biological creation, the most meaningful of moist mysteries among the great manifold of moist mysteries, is the construction of an organism from a single cell. Look at it backward so that things appear in reverse (I am giving you my own perspective): Viagra discarded, hair returned, skin tightened, that unfortunate marriage zipping backward, teeth uncapped, memories of a radiant young woman running through a field of lilacs, a bicycle with fat tires, skinned knees, Kool-aid and New Hampshire afternoons. But where memory fades in a glimpse of the noonday sun seen from a crib in winter, the biological drama only begins, for the rosy fat and cooing creature loitering at the beginning of the journey, whose existence I'm now inferring, the one improbably responding to *kitchy kitchy coo,* has come into the world as the result of a spectacular nine-month adventure, one beginning with a spot no larger than a pinhead and passing by means of repeated but controlled cellular divisions into an organism of ramified and intricately coordinated structures, these held together in systems, the systems in turn animated and controlled by a rich biochemical apparatus, the process of biological creation like no other seen anywhere in the universe, strange but disarmingly familiar, for when the details are stripped away, the revealed miracle seems cognate to miracles of a more familiar kind, as when something is read and understood.

The schedule by which this spectacular nine-month construction is orchestrated lies resident in DNA — and "schedule" is the appropriate word, for while the outcome of the drama is a surprise, the offspring proving to resemble his maternal uncle and his great-aunt (red hair, prominent ears), the process itself proceeds inexorably from one state to the next, and processes of this sort, which are combinatorial (cells divide), finite (it comes to an end in the noble and lovely creature answering to my name), and discrete (cells are cells), would seem to be essentially algorithmic in nature, the algorithm now making and marking its advent within the very bowels of life itself.

DNA is a double helix — this everyone now knows, the image as familiar as Marilyn Monroe — two separate strands linked to one another by a succession of steps, so that the molecule itself looks like an ordinary ladder seen under water, the strands themselves curved and waving. Information is stored on each strand by means of four bases — A, T, G, and C; these are by nature chemicals, but they function as symbols, the instruments by which a genetic message is conveyed.

A library is in place, one that stores information, and far away, where the organism itself carries on, one sees the purposes to which the information is put, an inaccessible algorithm ostensibly orchestrating the entire affair. Meaning is inscribed in molecules, and so there is something that reads and something that is read; but they are, those strings, richer by far than the richest of novels, for while Tolstoy's *Anna Karenina* can only suggest the woman, her black hair swept into a chignon, the same message carrying the same meaning, when read by the right biochemical agencies, can bring the woman to vibrant and complaining life, reading now restored to its rightful place as a supreme act of creation.

The mechanism is simple, lucid, compelling, extraordinary. In transcription, the molecule faces outward to control the proteins. In replication, it is the internal structure of DNA that conveys secrets, not from one molecule to another but from the past into the future. At some point in the life of a cell, double-stranded DNA is cleaved, so that instead of a single ladder, two separate strands may be found waving gently, like seaweed, the bond between base pairs broken. As in the ancient stories in which human beings originally were hermaphroditic, each strand finds itself longingly incomplete, its bases unsatisfied because unbound. In time, bases attract chemical antagonists from the ambient broth in which they are floating, so that if a single strand of DNA contains first A and then C, chemical activity (and chemical activity alone) prompts a vagrant T to migrate to A, and ditto for G, which moves to C, so that ultimately the single strand acquires its full complementary base pairs. Where there was only one strand of DNA, there are now two. Naked but alive, the molecule carries on the work of humping and slithering its way into the future.

A general biological property, intelligence is exhibited in varying degrees by everything that lives, and it is intelligence that immerses living creatures in time, allowing the cat and the cockroach alike to peep into the future and remember the past. The lowly paramecium is intelligent, learning gradually to respond to electrical shocks, this quite without a brain, let alone a nervous system. But like so many other psychological properties, intelligence remains elusive without an objective correlative, some public set of circumstances to which one can point with the intention of saying, There, that is what intelligence is or what intelligence is like.

The stony soil between mental and mathematical concepts is not usually thought efflorescent, but in the idea of an algorithm modern mathematics does offer an obliging witness to the very idea of intelligence. Like almost everything in mathematics, algorithms arise from an old wrinkled class of human artifacts, things so familiar in collective memory as to pass unnoticed. By now, the ideas elaborated by Gödel, Church, Turing, and Post have passed entirely into the body of mathematics, where themes and dreams and definitions are all immured, but the essential idea of an algorithm blazes forth from any digital computer, the unfolding of genius having passed inexorably from Gödel's incompleteness theorem to Space Invaders VII rattling on an arcade Atari, a progression suggesting something both melancholy and exuberant about our culture.

The computer is a machine, and so belongs to the class of things in nature that do something; but the computer is also a device dividing itself into aspects, symbols set into software to the left, the hardware needed to read, store, and manipulate the software to the right. This division of labor is unique among man-made artifacts: it suggests the mind immersed within the brain, the soul within the body, the presence anywhere of spirit in matter. An algorithm is thus an ambidextrous artifact, residing at the heart of both artificial and human intelligence. Computer science and the computational theory of mind appeal to precisely the same garden of branching forks to explain what computers do or what men can do or what in the tide of time they have done.

Molecular biology has revealed that whatever else it may be, a living creature is also a combinatorial system, its organization controlled by a strange, hidden, and obscure text, one written in a biochemical code. It is an algorithm that lies at the humming heart of life, ferrying information from one set of symbols (the nucleic acids) to another (the proteins).

The complexity of human artifacts, the things that human beings make, finds its explanation in human intelligence. The intelligence responsible for the construction of complex artifacts — watches, computers, military campaigns, federal budgets, this very essay — finds its explanation in biology. Yet however invigorating it is to see the algorithmic pattern appear and reappear, especially on the molecular biological level, it is important to remember, if only because it is so often forgotten, that in very large measure we have no idea how the pattern is amplified. Yet the explanation of complexity that biology affords is largely ceremonial. At the very heart of molecular biology, a great mystery is vividly in evidence, as those symbolic forms bring an organism into existence, control its morphology and development, and slip a copy of themselves into the future.

The transaction hides a process never seen among purely physical objects, one that is characteristic of the world where computers hum and human beings attend to one another. In that world intelligence is always relative to intelligence itself, systems of symbols gaining their point from having their point gained. This is not a paradox. It is simply the way things are. Two hundred years ago the French biologist Charles Bonnet asked for an account of the "mechanics which will preside over the formation of a brain, a heart, a lung, and so many other organs." No account in terms of mechanics is yet available. Information passes from the genome to the organism. Something is given and something read; something ordered and something done. But just who is doing the reading and who is executing the orders, this remains unclear.

MARK CHERRINGTON

To Save a Watering Hole

FROM *Discover*

ONE CLOUDLESS JULY AFTERNOON in 1995, Dalit Yosef and her nine-year-old son Erez left their home in Eilat, Israel, and set off for a nearby bird sanctuary. Their route took them down the hill through the bustling city, past brown stucco houses and apartment blocks, down streets lined with palms and tropical flowers. To the east, across the Red Sea, the Jordanian mountains rose in a stark palisade above Aqaba; to the south the desert mountains of Saudi Arabia were a dark shadow on the horizon. Yet Dalit took no comfort in the view. That idyllic locale had become an environmental battlefield.

On one side lay a legion of local developers, bent on turning Eilat into Israel's premier resort town. On the other side was Dalit's husband, ornithologist Reuven Yosef, equally determined to set aside part of the landscape for wildlife. The sanctuary that Dalit and Erez were visiting that day was ground zero. Economically, it was quite valuable: Among other things, the mayor had suggested turning it into a motocross track or a zoo. But biologically it was priceless. Every spring and autumn, a large part of Europe and Asia's migrating bird population — 1.5 billion birds in all — passed through Eilat on the way to and from Africa. They stopped because the sanctuary was on the only land bridge between Africa, Asia, and Europe, and because it offered them the only food for hundreds of miles in any direction. Without the sanctuary, millions of the birds might die, and without those birds, ecosystems on three continents would be threatened.

As director of the International Birding and Research Center in town, Reuven Yosef was attacking Eilat's developers publicly and relentlessly, fighting to keep bulldozers away from the sanctuary. But his opponents were just as vehement. For the past year, people had been calling the Yosefs' house anonymously, making threats. "If you don't leave Eilat," they would say, "something really terrible is going to happen to your family." First one and then another of Reuven's jeeps was vandalized. At the sanctuary, fences were trampled, doors broken in, seedlings uprooted, and field equipment sabotaged so often that Yosef was spending 10 percent of his budget on replacement equipment. In each case, the culprits were never identified, and the police wrote off the attacks as "incidents of no public interest."

There was worse to come. When Dalit and her son reached the sanctuary that day in July, they walked through the trees to Reuven's research station, expecting to see the family dog, Jenny. Instead, as Dalit rounded the corner, she stopped in her tracks and screamed, spinning Erez around and covering his eyes. Jenny was hanging from her chain, which had been tied to the roof of the building, with a piece of paper tacked to her dead body. The words on the paper said, "Get Out!"

Four years later, describing the incident, Reuven Yosef's voice catches, then hardens. The pace of development has quickened, he says, and the attacks have escalated: A year after Jenny was killed, Yosef went to the sanctuary to find his research station burned to the ground. Yet he hasn't budged an inch. He is not a large man, but he is stocky and swarthy and black-bearded and radiates defiance. In his normal stance — feet planted apart and arms folded over his chest — he wouldn't be out of place on the prow of a pirate ship. His black eyes are unflinching, and when he locks them on you, your instinct is to take a step back. His physical attitude suggests a falcon: compact, alert, fearless, and always prepared to hurtle headlong at his prey.

Yosef's zeal is equal parts personal and scientific, a function of both his upbringing and the gravity of the situation in Eilat. The son of a fighter pilot, educated in a military boarding school, he left his native India in 1974 at age sixteen, after the Yom Kippur War. "Being an Indian Jew, I was sort of brainwashed in the Zionist ideal that living in Israel is the thing for a Jew," he says, "so I wanted to come and give whatever I could." He soon landed in one of the Israeli army's elite units and was promoted to a high rank over the course of several extremely dangerous missions. But even then his loyalties were divided.

One day, during an excursion behind enemy lines, Yosef's unit came upon a stork riddled with shrapnel. Yosef wanted to evacuate it, but bullets and bombs were flying all around. So he and the unit's doctor huddled by a rock and worked for hours removing the shrapnel and dressing the wounds. From then on, Yosef says, "My unit knew that any animal that we found wounded in the field was treated. If we thought the animal could survive in the field, we released it, and if not, we brought it back and rehabilitated it." His men didn't necessarily agree with his approach, he says: "They humored me."

Yosef's compassion for wildlife — cultivated, he says, by an uncle who was an eminent zoologist — grew like a raging fire. "I realized back then the kinds of things that were plowed under in the name of human requirements, of Zionist ideals." After leaving the army in 1979, he got degrees in biology and education at the University of Haifa, earned his doctorate in zoology at Ohio State University, and went on to do research at Cornell under renowned biologist Tom Eisner. By the time he became director of the center in Eilat in 1993, he was equally adept at research and warfare: the perfect man for the job.

Walking around the sanctuary that would become the center of Yosef's life — and nearly the cause of his death — one might wonder what all the fuss is about. The place is pretty enough. Tall, drooping jujube, or Christ's-thorn, trees shade bushes of sea blite, the primary plant for the birds, and dozens of other species. Two large ponds are dotted with rocks and islands, and the islands are covered with more sea blite. But compared with other exotic and embattled wildlife areas like the Amazon, Madagascar, and the Serengeti, Eilat's sanctuary seems unremarkable. What could be so critical about it?

The answer lies in Yosef's stunning statistics. Every spring, he estimates, 3 million raptors pass over Israel, including 800,000 honey buzzards, 65,000 Levant sparrow hawks, 460,000 steppe buzzards, 142,000 lesser spotted eagles, and 30,000 steppe eagles

— and that count doesn't include 1.2 billion songbirds and waders. They come here from the Sahel, the Sahara, and the Sinai, where the desert is so dry people have died of dehydration in eight hours. The birds come by the tens of thousands, great swirling clouds of them, sunbaked and exhausted, desperate to reach this piece of land. Even under the best of circumstances, 60 percent of some migrating species may die en route, and some survivors arrive with all their fat and most of their muscle metabolized. For birds flying south, the sanctuary is a last, critical pit stop. For birds flying north, it's a lifesaving oasis.

But if Eilat's sanctuary were lost, the real victims would be elsewhere. The birds that depend on this sanctuary play crucial ecological roles in Europe, Asia, and Africa, distributing seeds of fruit-bearing plants, pollinating flowers, and controlling insect and rodent populations. In 1921, American ornithologist Edward Forbush calculated that birds in the United States reduce forest and agricultural pests by 28 percent, saving the country an estimated $444 million in crop and timber losses. And that figure is in 1921 dollars. Today birds in Europe, Asia, and Africa are worth billions of dollars to farmers.

It's hard, however, to put a precise figure on the value of the sanctuary. "This subject is a minefield," Yosef says. "There are those waiting for me to make claims and then to slam me with them. But we have established that more than ninety passerine species stage at Eilat. If we take into account that several raptor, wader, and pelagic species also stop over here, the numbers are a considerable proportion of Eurasia's breeding migratory population. Maybe even several tens of percentage."

For all his caution in making estimates, Yosef has already witnessed what will happen if this sanctuary is lost. When he first came to Eilat, the marsh that birds once used here was being built over. "You would see birds wandering around town, exhausted and looking for food," he says, "but all that was planted were exotic species." Some of the birds were feeding in local, pesticide-laden farmlands.

There wasn't much point in directing a birding center if it meant watching birds die, so Yosef threw himself behind a local initiative to build a bird sanctuary. A local bird lover and former deputy mayor named Shmulik Tagar had scouted out a possible location for the sanctuary at an abandoned 160-acre landfill on the edge of town. Yosef helped Tagar persuade the town to donate the land, then the two men talked the town's developers into taking the topsoil they were excavating and dumping it onto the landfill. Rather than fight for a share of Eilat's precious water supply, the activists went after the town's sewage, then being dumped raw into the Red Sea. They knew that the sewage, in time, would kill one of the city's major sources of income: its spectacular coral reef. So they helped organize demonstrations and bumper-sticker promotions. After three years, the town agreed to build a treatment plant just above Yosef's land.

The rest was easy by comparison. Using a backhoe, Yosef dug two ponds and a drip-irrigation network. Then he raised money to buy plastic tubing, set up a solar pump, and pulled down some of the nutrient-rich treated water from the sewage canal that ran alongside his land. Finally, he convinced the Jewish National Fund to help him seed all the native plants that had grown in the original salt marsh. Today his garbage dump is an ornithological paradise. Only one twentieth the size of the original marsh, it supports a far greater density of plants, and even more food for the birds, thanks to irrigation.

In most places, such an achievement would make Reuven Yosef a hero. But Eilat is a one-business town, and that business is tourism. Over the past five years, the overall number of hotel rooms has doubled, to 10,000, and the population has leaped from 45,000 to 90,000. There are millions to be made developing the desert, and 160 acres of natural habitat in the heart of it seems to many like land wasted.

Driving through town one afternoon, Yosef launches into what must be a well-worn tirade against the appalling environmental cost of Eilat's boomtown mentality. Those flowers and waving palms along the streets are all exotic species, he says. They are displacing native plants and animals. Those glass-bottomed boats are destroying the coral on which they make a living. "When I told them they had to think of future generations," he says, "they told me, 'Let them worry about their own problems; I'm going to make my money now.'"

Yosef's opponents, for their part, make no apologies. "When I was born, in Tel Aviv, there were fewer than 1 million people here," says Rina Maor, the director for southern Israel at the Ministry of Tourism. "And now we are 6 million. Yes, we pollute, we do. What can we do? We don't live in tents. I know that we disturb not only the birds but also the corals and fish, and we are not so nice, we human beings — we are sometimes very cruel. But what can we do?"

Yosef threatens much more than a few hotel builders, it seems; he challenges the very idea of Israel. More than fifty years ago, the country's founder and first prime minister, David Ben-Gurion, declared that Israel's mission was "to make the Middle East bloom again." As a Jewish nation, Ben-Gurion believed, Israel would gain credibility only if Jews established an undeniable presence. In Israel, making the desert bloom has been both a political and a cultural obligation.

On paper, Israel appears to be protecting its environment: Nearly a quarter of its land is set aside as national parks or nature reserves. But 63 percent of those reserves are less than half a square mile in area. Moreover, half of the official preserves are on land used by the military. The Ministry of the Environment can create a reserve only if the Ministry of Agriculture or the Ministry of Defense approves. Finally, because the government technically owns all the land in Israel, any given piece can be confiscated and reclassified at any time.

"Israel's got this paradox," says Alon Tal, who has just finished a book on Israeli environmental history. "We have remarkable biodiversity for a country this size because of our location on three continents. I mean, in my backyard we've got hyenas living with wolves and foxes, all in the middle of the desert. The trouble is, our population is growing very fast and the standard of living is growing very fast. We've got a developing country's demographic profile with a Western country's economic profile. And that can lead to a lot of problems."

Last December a government official charged Yosef with being anti-Israeli. Although he was eventually cleared in a hearing, after dozens of scientists and friends from around the world came to his defense, his accuser had no lack of allies. Just this past winter, Yosef mounted the barricades again when the Israeli government, as part of a peace agreement with Jordan, proposed a new international airport not far from his sanctuary. Arguing that the region was of unique biological value, Yosef and a group of other environmentalists petitioned for a change in the plan. The airport would disrupt the birds' migration, they argued, and Israel had an obligation not to let that happen: Europe, Asia, and Africa's birds depended on it.

Somehow, it worked: Yosef and his colleagues persuaded the minister of the environment to take up their cause. Only a single terminal will be built in Israel, and that on disturbed land away from the sanctuary. Meanwhile, the world environmental community is finally beginning to recognize Yosef's efforts. Next month, Yosef will be given the prestigious Rolex award for conservation.

Still, he is hardly ready to let down his guard. Walking through the sanctuary one day, he proudly points out his new research station. When the original structure was burned down, he says, he gathered the building's charred timbers and turned them into seating for sanctuary visitors. Then he launched a fund-raising campaign to rebuild it. The new structure is made of chemically treated fire-resistant wood. It's set on a concrete slab and equipped with locking shutters for the windows. "If they want to get rid of this one," Yosef says, "they'll have to drop a bomb on it." His new dog, though, stays at home.

It's January now — the calm before the storm of migration — but scanning the sky, Yosef notices something flying against the Jordanian mountains in the distance. "Oh, look," he says. "There are some ducks coming in. Looks like mallards, or maybe pintails. They're early this year." Even with binoculars it's impossible to see them. Then the birds pass in front of the mountains and against the clear sky. To most people, they would look like grains of pepper at the far end of a football field. But Yosef has uncommonly good vision: Where anyone else would see only desert, he sees birds.

EDWIN DOBB

New Life in a Death Trap

FROM *Discover*

PITY THE SNOW GEESE that settled on Lake Berkeley as a stopover one stormy night in November 1995. The vast lake, covering almost 700 acres of a former open-pit copper mine in Butte, Montana, holds some 30 billion gallons of highly acidic, metal-laden water — scarcely a suitable refuge for migrating birds stalled by harsh weather. So when the flock rose up and turned southward the following morning, almost 350 carcasses were left behind. Autopsies showed their insides were lined with burns and festering sores from exposure to high concentrations of copper, cadmium, and arsenic.

Today one need only stand on the viewing platform and look at the pit — the lifeless yellow and gray walls that stretch for a mile in one direction and a mile and a half in the other and the dark, eerily placid lake — to see that it's hostile toward living things. Surely nothing could survive these perilous waters. But in 1995, the same year the birds died, a chemist studying lake composition retrieved some rope coated with brilliant green slime and took it to his colleagues at Montana Tech of the University of Montana, an institution locals proudly call the Tech. Having evolved in partnership with one of the world's richest and longest-running mining districts, it remains a world-class engineering and mining school. Grant Mitman, one of just three full-time biologists on the faculty, quickly identified the slime as a robust sample of single-celled algae known as *Euglena mutabilis*. Life had somehow established an outpost in the liquid barren that is the Berkeley Pit.

For Mitman, finding *Euglena* proved uncannily fortunate. At

Dalhousie University in Halifax, Nova Scotia, where he received his doctorate, his passion was algae. "I trained all my life to be a marine biologist," he says, noting the irony in then having taken a post at an engineering school in the Rocky Mountains. Just as a landlocked, man-made toxic lake has reunited this scientist and his favorite subject, so too has it galvanized the long-standing interest of chemists Don and Andrea Stierle, a husband-and-wife team who also work at the university. The Stierles have spent their lives searching for naturally occurring compounds that can be used in agriculture and medicine. For them, the menagerie of small organisms — more than forty — discovered in Lake Berkeley during the past five years holds much potential.

Even more important, perhaps, is the promise that some of those organisms can be employed to reclaim the lake — and other similar repositories of mine wastewater — by neutralizing acidity and absorbing dissolved metals. Beyond these potential benefits are possible theoretical advances in biology. Each new discovery of a so-called extremophile — an organism adapted to unusually harsh conditions — helps illuminate fundamental biological processes, from metabolic dynamics to the means and course of evolution, both here on Earth and elsewhere in the universe.

Lake Berkeley was born of human appetite and geological happenstance. During the early 1880s, just as electricity was lighting up cities and the need for copper mushroomed, an ambitious prospector named Marcus Daly discovered an enormous deposit of the red metal 300 feet down in his own Anaconda Mine. For the next fifty years, Butte provided a third of the copper used in the United States and a sixth of the world's supply — all from a mining district only four miles square. Thereafter the "Richest Hill on Earth," as journalists often referred to the place, continued to yield vast amounts of metals.

After the Second World War, the shafts grew deeper — one of them eventually reached a level one mile beneath the surface — but the quality of the ore diminished. Mining officials decided to switch from labor-intensive and dangerous underground operations to open-pit mining, a more efficient method for extracting low-grade ore. Excavation began in 1955, and soon the pit became the world's largest truck-operated mine, along the way displacing some Italian and Serbo-Croatian neighborhoods that had grown up around the original mines on the east end of town. Mining came to a halt in the early 1980s, as did the pumps that had been sucking groundwater out of the mines for a century. The flooding began.

Stroll across the mining landscape of Butte today and you will discover why the water has had such a profound environmental effect. The land is dull ocher, and the air smells like rotting eggs. If you look closely at the waste rock, you will see pyrite crystals — fool's gold — everywhere. These are all signs of sulfur. The bedrock is shot through with it. When exposed to air and water, long-buried sulfide minerals produce sulfuric acid, which also helps dissolve other minerals from surrounding rock. Acid-tolerant bacteria that thrive on iron and sulfur compounds hasten this process, and when the pumps were shut down, the Berkeley Pit became an immense chemical transformer producing ever-greater amounts of toxic soup. Making matters worse, it's self-perpetuating. By all accounts, groundwater will continue to migrate into Lake Berkeley indefinitely. Because of this threat to the community, the Environmental Protection Agency added the pit to the federal Superfund list in 1987. The designation also made it part of the country's largest complex of Superfund sites — a series that includes a good part of Butte and the upper 120 miles of the Clark Fork River watershed.

Today Lake Berkeley is the country's largest and most unusual body of contaminated water. With a pH of 2.6, it's as acidic as cola or lemon juice. Besides copper, cadmium, and arsenic, the water contains a dozen other metals, including aluminum, iron, manganese, and zinc. But it is precisely because of these harsh conditions that the lake has caught the attention of life scientists. "We divide the organisms we've found into two categories," says Andrea Stierle. Dressed casually in a T-shirt, jeans, and sneakers, the chemist stands in her lab next to a counter covered with petri dishes and Erlenmeyer flasks, each one containing a brightly colored fungus culture. "The first group we call survivors," she explains. "They don't really like the environment, but they put up with it. They're able to defend themselves."

Less numerous but far more interesting to Stierle and her husband are the dynamic lake inhabitants they call thrivers. Like the survivors, these organisms arrive by accident — transported by wind and runoff, deposited by birds, sloughed off boat bottoms or old mine timbers. But unlike their less-prepared counterparts, they actually flourish in the presence of acidity and make use of some of the dissolved metals in the lake. A toxic-waste dump is a biological haven to thrivers. As they reproduce — fungi require only a week to do so, bacteria but a day — characteristics that render them more fit become widespread. Metabolic processes are affected. Or as Andrea Stierle puts it, "New environmental niches mean new microbes, new microbes mean new chemistry, and new chemistry means new chemical compounds."

The Stierles have good reason to believe that the microbes in Lake Berkeley are a likely source of useful chemicals. Natural compounds have preoccupied them for twenty years, and their partnership has yielded several notable discoveries. The one that makes them most proud occurred in the early 1990s. Then, research showed that a substance called taxol is an effective agent in the treatment of breast and ovarian cancer. In a third of the women receiving taxol, tumors actually shrank. But the news was bittersweet. Taxol comes from the bark of Pacific yew trees, a species native to the Pacific Northwest but nearly extinct. "Ninety-five percent of them had been cut down or burned as slash," Andrea Stierle says. The few trees left couldn't provide enough taxol to meet demand.

While everyone else concentrated on synthesizing the substance and developing methods for growing yew trees more rapidly, she and Don followed a tactic called "biorational serendipity" — a combination of scientific deduction and clever, if sometimes prolonged, sleuthing. Taxol might be in yew bark, they reasoned, because a parasitic or symbiotic microbe manufactures it there. "Life is everywhere" Andrea says, "and all kinds of bacteria and fungi are found on plants." Often they produce compounds that have never before been seen or exploited. The best-known example, of course, is penicillin, which was first extracted from a mold.

For almost two years, Andrea and Don crisscrossed the Pacific Northwest, taking bark samples and, along with plant pathologist Gary Strobel of Montana State University, testing for the presence of taxol. Finally, in 1992, in bark from a yew tree in Glacier Park, they found what they were looking for — a previously unknown fungus that produces the cancer-killing substance. In honor of Andrea, they named the new organism *Taxomyces andreanae* and applied for a patent. Five years later, Bristol-Myers Squibb purchased the commercial rights.

The Stierles are using much the same reasoning to study the biota of the Berkeley Pit. "Whether in defense or offense, every microbe uses its chemistry to protect itself," Andrea Stierle explains. In other words, bacteria, fungi, and the like manufacture substances that can be poisonous to other microbes. The generic term for such chemicals is "secondary metabolites" — unique compounds that organisms assemble from the basic building blocks, or primary metabolites, such as carbon or hydrogen, that most living things hold in common. "It's among the secondary metabolites," Stierle says, "that we find natural products that can benefit medicine or agriculture."

After finding a promising secondary metabolite, the Stierles use standard bioassays to tell whether and to what degree it is toxic. The first is called the brine shrimp lethality test. "It's been found that compounds that kill brine shrimp are more likely to destroy cancer cells," Stierle says. A second test involves *E. coli,* a common intestinal bacterium. If the compound repairs DNA in a damaged *E. coli,* it might also work against cancer. In another assay, the compound is applied along with *Agrobacterium tumifaciens,* a tumor-causing microbe, to potato slices. Previous research has shown that if the new chemical protects the potato against tumor formation, it could prove useful in medicine as well. So far the Stierles have isolated five novel compounds, all from a single fungus. Each one is lethal to brine shrimp. "We've sent them to the National Cancer Institute for further study," Andrea says.

Because of their success with biorational serendipity, the Stierles fully expect to discover many more new substances among the other fungi and bacteria from Lake Berkeley. "We're taking samples everywhere," Stierle says, "from the surface, the entire 700-foot column of water, the sediments at the bottom." And as time goes by, the odds improve, because with time the thrivers are more likely to undergo change. "Under such hostile conditions the pressure to mutate is intense," Stierle says. "In fact, we may already be seeing the results of natural selection." Clearly the mine flood was an environmental disaster with potentially deadly consequences for snow geese and other creatures, but it is now — eighteen years and thousands of microbial generations later — proving an engine of evolution.

While the Stierles watch expectantly, confident that contingency will yield a chemical bounty, biologist Grant Mitman is preparing a recipe for directing the community of life in the Berkeley Pit. Some 3 million gallons of groundwater seep into the lake daily, raising the surface by about one foot a month. Engineers predict that in about twenty years the water in the pit will rise to the same level as the surrounding groundwater. From that point on, any more water that enters the ground will flow in the opposite direction, polluting the alluvial aquifer in the valley below the mine and discharging toxic metals into Silver Bow Creek, the headwaters of the Clark Fork River. To prevent this calamity, the Atlantic Richfield Company, which is responsible for Superfund reclamation costs, has to construct a treatment plant before the critical level is reached. But the process under consideration — treating the water with lime, to which metals naturally bind — would produce between 500 and 1,000 tons of toxic sludge each day. Like many others, Mitman believes there is a better way. His way features his favorite microbes.

"I'm looking for organisms that will clean up the water," Mitman explains. "And I believe algae are the best candidates." Slender, tall, with wire-rim glasses, the forty-two-year-old biologist manages to appear professorial even as he waxes algal while standing in his walk-in environmental chamber at Montana Tech, holding a flask of light green water in one hand and one of dark green water in the other. The sign on the outside of the room reads "Growth Chamber." It is here that Mitman induces miniature blooms in water taken from the contaminated lake. The emerald bloom is *Euglena mutabilis,* the first new resident to be identified; its darker counterpart is *Chlorella ellipsoida vulgaris,* one of four other algae Mitman has isolated. Holding up the flask of *Chlorella,* he says with unmistakable optimism, "This is what the Berkeley Pit could look like someday."

That green should be the color of salvation might be fitting in a place where so much was sacrificed in the name of industrialization. But behind the symbolism is a compelling biological argument. Certain algae consume metals. Others produce bicarbonate, which reduces acidity. The right organisms in the right numbers, the logic goes, would help remedy the two most noxious features of Lake Berkeley. But that is not the only potential benefit. Algae also convert sunlight, carbon dioxide, and water into sugar. And sugar, Mitman says, "is what makes any system come alive." It's the food that other, larger organisms, such as protozoans and fungi, need to survive. Some of these larger microbes also reduce acidity. But most important, they concentrate metals tenfold whenever they consume metal-eating algae, a process sometimes referred to as biological magnification. And when an organism dies, it drifts into sediments at the bottom of the lake, where any metals it might contain are impounded. "The key," Mitman says, "is to get the algae going first."

That's just what he is doing on a small scale in his laboratory. Under the auspices of Montana Tech's Mine Waste Technology Program, Mitman is systematically concocting brews of Berkeley algae. He varies such factors as light and temperature, but he's most interested in what nutrients each batch receives. Unlike bacteria, protozoans, and fungi, algae feed on fairly inexpensive and widely available inorganic nutrients, such as nitrogen and phosphorus. By doing no more than adding these chemicals, Mitman has been able to trigger an extraordinarily rapid growth of algal colonies. Ironically, *Euglena,* the organism that launched the current research programs of both the Stierles and Mitman, turns out to be highly resistant to metals. It actively excludes them, flourishing in their presence without making use of them. "We even grew one sample on a piece of solid copper," Mitman explains.

Chlorella has proved more promising. In initial tests, it reduced the mineral content of the pit water by as much as 10 percent. That may not seem like much, but *Chlorella* is only one of several indigenous organisms that in all likelihood can reduce the lake's toxic contents. And as Mitman says, "Every grain of metal that can be removed will save lots of money in the long run." He is now focusing on another denizen, *Chromulina freiburgensis,* an alga that has already been shown to concentrate metals in other settings but had never been seen before in acidic mine water. Following the lab work will come field tests. Mitman envisions barrels floating on Lake Berkeley — each housing an experimental brew of ordinary lake water, algae, and various nutrients.

Mitman is convinced that inorganic nutrients can gradually bring about the natural recovery of Lake Berkeley. Creating a big algal bloom could be as simple as spreading nitrate across the surface of the water, or some form of mixing may be the best approach. Some nitrogen-fixing bacteria inhabit the upper levels of the lake, he explains, and when they extract nitrogen from the air for their metabolic needs, they process nitrogen that other organisms in the water can use. In large enough numbers, they could supply the additional nutrients needed to make Lake Berkeley continue blooming on its own. "Eventually the system could be self-sustaining," Mitman says. Self-sustaining and ever paradoxical. Whether brownish red due to high iron-sulfide content or black because of metal-concentrating algae, Lake Berkeley will remain a fascinating, if forbidding, sight — a testament to nature's resiliency as well as a sobering reminder of the extremes we will go to get the resources we want. Life, as Andrea Stierle says, may be everywhere. But it is not everywhere guaranteed.

GREGG EASTERBROOK

Abortion and Brain Waves

FROM *The New Republic*

NO OTHER ISSUE in American politics stands at such an impasse. Decades after *Roe* v. *Wade,* the abortion debate remains a clash of absolutes: one side insists that all abortions be permitted, the other that all be prohibited. The stalemate has many and familiar causes, but a critical and little-noticed one is this: public understanding has not kept pace with scientific discovery. When *Roe* was decided in 1973, medical knowledge of the physiology and neurology of the fetus was surprisingly scant. Law and religion defined our understanding, because science had little to say. That is now changing, and it is time for the abortion debate to change in response.

Quietly, without fanfare, researchers have been learning about the gestational phases of human life, and the new information fits neither the standard pro-choice position nor the standard pro-life position. As far as science can tell, what happens early in the womb looks increasingly like cold-hearted chemistry, with the natural termination of potential life far more common than previously assumed. But science also shows that by the third trimester the fetus has become much more human than once thought — exhibiting, in particular, full brain activity. In short, new fetal research argues for keeping abortion legal in the first two trimesters of pregnancy and prohibiting it in the third.

This is a message neither side wants to hear. But, as the Supreme Court prepares to take up the abortion issue for the first time in nearly a decade, new fetal science may provide a rational, nonideological foundation on which to ground the abortion compromise that currently proves so elusive. And, curiously enough, by supporting abortion choice early in pregnancy while arguing against it later on, the science brings us full circle — to the forgotten original reasoning behind *Roe.* Many religious interpretations today hold that life begins when sperm meets egg. But this has not always been so; until 1869, for example, the Catholic Church maintained that life commenced forty days after conception. Derived from interpretation rather than from Scripture — the Bible says nothing about when the spark of life is struck — the notion that sacredness begins when sperm meets egg hinges on the assumption that it is God's plan that each act of conception should lead to a baby.

But new science shows that conception usually does not produce a baby. "The majority of cases in which there is a fertilized egg result in the nonrealization of a person," says Dr. Machelle Seibel, a reproductive endocrinologist at the Boston University School of Medicine. What exists just after conception is called a zygote. Research now suggests that only about half of all zygotes implant in the uterine wall and become embryos; the others fail to continue dividing and expire. Of those embryos that do trigger pregnancy, only around 65 percent lead to live births, even with the best prenatal care. The rest are lost to natural miscarriage. All told, only about one-third of sperm-egg unions result in babies, even when abortion is not a factor.

This new knowledge bears particularly on such controversies as the availability of "morning after" birth-control pills, which some pharmacy chains will not stock. "Morning after" pills prevent a zygote from implanting in the uterine wall. If half of all fertilized eggs naturally do not implant in the uterine wall, it is hard to see why a woman should not be allowed to produce the same effect using artificial means.

More generally, the evidence that two-thirds of conceptions fail regardless of abortion provides a powerful new argument in favor of choice in the early trimesters. Perhaps it is possible that God ordains, for reasons we cannot know, that vast numbers of souls be created at conception and then naturally denied the chance to become babies. But science's new understanding of the tenuous link between conception and birth makes a strong case that what happens early in pregnancy is not yet life in the constitutional sense.

Yet, if new science buttresses the pro-choice position in the initial trimesters, at the other end of pregnancy it delivers the opposite message. Over the past two decades it has become increasingly clear that by the third trimester many fetuses are able to live outside the mother, passing a basic test of personhood. Now research is beginning to show that by the beginning of the third trimester the fetus has sensations and brain activity and exhibits other signs of formed humanity.

Until recently most physicians scoffed at the idea of fetal "sentience." Even newborns were considered incapable of meaningful sensation: until this generation, many doctors assumed that it would be days or weeks before a newly delivered baby could feel pain. That view has been reversed, with the medical establishment now convinced that newborns experience complex sensations. The same thinking is being extended backward to the third-trimester fetus.

Over the past decade, pediatric surgeons have learned to conduct within-womb operations on late-term fetuses with correctable congenital conditions. As they operated within the womb, doctors found that the fetus is aware of touch, responds to sound, shows a hormonal stress reaction, and exhibits other qualities associated with mental awareness. "The idea that the late-term fetus cannot feel or sense has been overturned by the last fifteen years of research," says Dr. Nicholas Fisk, a professor of obstetrics at the Imperial College School of Medicine in London.

Most striking are electroencephalogram (EEG) readings of the brain waves of the third-trimester fetus. Until recently, little was known about fetal brain activity because EEG devices do not work unless electrodes are attached to the scalp, which is never done while the fetus is in the womb. But the past decade has seen a fantastic increase in doctors' ability to save babies born prematurely. That in turn has provided a supply of fetal-aged subjects who are out of the womb and in the neonatal intensive care ward, where their EEG readings can be obtained.

EEGs show that third-trimester babies display complex brain activity similar to that found in full-term newborns. The legal and moral implications of this new evidence are enormous. After all, society increasingly uses cessation of brain activity to define when life ends. Why not use the onset of brain activity to define when life begins?

Here is the developmental sequence of human life as suggested by the latest research. After sperm meets egg, the cells spend about a week differentiating and dividing into a zygote. One to two weeks later the zygote implants in the uterine wall, commencing the pregnancy. It is during this initial period that about half of the "conceived" sperm-egg pairings die naturally. Why this happens is not well understood: one guess is that genetic copying errors occur during the incipient stages of cellular division.

The zygotes that do implant soon transform into embryos. During its early growth, an embryo is sufficiently undifferentiated that it is impossible to distinguish which tissue will end up as part of the new life and which will be discarded as placenta. By about the sixth week the embryo gives way to the fetus, which has a recognizable human shape. (It was during the embryo–fetus transition, Augustine believed, that the soul is acquired, and this was Catholic doctrine for most of the period from the fifth century until 1869.) Also around the sixth week, faint electrical activity can be detected from the fetal nervous system. Some pro-life commentators say this means that brain activity begins during the sixth week, but, according to Dr. Martha Herbert, a neurologist at Massachusetts General Hospital, there is little research to support that claim. Most neurologists assume that electrical activity in the first trimester represents random neuron firings as nerves connect — basically, tiny spasms.

The fetus's heart begins to beat, and by about the twentieth week the fetus can kick. Kicking is probably a spasm, too, at least initially, because the fetal cerebral cortex, the center of voluntary brain function, is not yet "wired," its neurons still nonfunctional. (Readings from twenty- to twenty-two-week-old premature babies who died at birth show only very feeble EEG signals.) From the twenty-second week to the twenty-fourth week, connections start to be established between the cortex and the thalamus, the part of the brain that translates thoughts into nervous-system commands. Fetal consciousness seems physically "impossible" before these connections form, says Fisk, of the Imperial College School of Medicine.

At about the twenty-third week the lungs become able to function, and, as a result, twenty-three weeks is the earliest date at which premature babies have survived. At twenty-four weeks the third trimester begins, and at about this time, as the cerebral cortex becomes "wired," fetal EEG readings begin to look more and more like those of a newborn. It may be a logical consequence, either of natural selection or of divine creation, that fetal higher brain activity begins at about the time when life outside the mother becomes possible. After all, without brain function, prematurely born fetuses would lack elementary survival skills, such as the ability to root for nourishment.

At about twenty-six weeks the cell structure of the fetal brain begins to resemble a newborn's, though many changes remain in store. By the twenty-seventh week, according to Dr. Phillip Pearl, a pediatric neurologist at Children's Hospital in Washington, D.C., the fetal EEG reading shows well-organized activity that partly overlaps with the brain activity of adults, although the patterns are far from mature and will continue to change for many weeks. By the thirty-second week, the fetal brain pattern is close to identical to that of a full-term baby.

Summing up, Paul Grobstein, a professor of neurology at Bryn Mawr University, notes, "I think it can be comfortably said that by the late term the brain of the fetus is responding to inputs and generating its own output. The brain by then is reasonably well developed. But we still don't know what within the fetal brain corresponds to the kind of awareness and experience that you and I have." The fetus may not know it is a baby or have the language-ordered thoughts of adults. But Grobstein points out that from the moment in the third trimester that the brain starts running, the fetus can experience the self/other perceptions that form the basis of human consciousness — since the womb, to it, represents the outside world.

In 1997, the Royal College of Obstetricians and Gynecologists, Britain's equivalent to a panel of the National Academy of Sciences, recommended that, because new research shows that the fetus has complex brain activity from the third trimester on, "practitioners who undertake termination of pregnancy at twenty-four weeks or later should consider the requirements for feticide or fetal analgesia and sedation." In this usage, "feticide" means killing the fetus the day before the abortion with an injection of potassium that stops the fetus's heart, so that death comes within the womb. Otherwise, the Royal College suggests that doctors anesthetize the fetus before a third-trimester termination — because the fetus will feel the pain of death and may even, in some sense, be aware that it is being killed.

If a woman's life is imperiled, sacrificing a third-trimester fetus may be unavoidable. But the American Medical Association (AMA) says late-term abortions to save the mother's life are required only under "extraordinary circumstances"; almost all late-term abortions are elective. In turn, the best estimates suggest that about 750 late-term abortions occur annually in the United States, less than 1 percent of total abortions. (An estimated 89 percent of U.S. abortions occur in the first trimester, ethically the least perilous time.) Pro-choice advocates sometimes claim that, because less than 1 percent of abortions are late-term, the issue doesn't matter. But moral dilemmas are not attenuated by percentages: no one would claim that 750 avoidable deaths of adults did not matter.

On paper the whole issue would seem moot, because Supreme Court decisions appear to outlaw late-term abortion except when the woman's life is imperiled. But in practice the current legal regime allows almost any abortion at any time, which turns out to be a corruption of *Roe.*

In its 1973 opinions in *Roe* and a companion case called *Bolton,* the Supreme Court established an abortion hierarchy: during the first trimester, there would be essentially no restrictions; during the second trimester, states could regulate abortion, but only to insure that procedures were carried out by qualified practitioners; during the third trimester, states could prohibit abortion, except when necessary "to preserve the life or health of the mother." (In abortion law, the Supreme Court sets ground rules, but states enact the regulations; Congress can sometimes intervene.) *Roe*'s third-trimester standard was considered largely theoretical, because in 1973 doctors were generally unable to perform safe late-term abortions. That would change.

Roe was premised on the idea that the Constitution protects medical privacy, an important concept in law for everyone, not just women. But even constitutional rights may be regulated, as, for example, libel laws regulate free speech. *Roe* did not grant an unqualified privilege: it held that a woman's claim to make her own medical choices is strong in the first trimester of pregnancy, moderate in the second, and weak in the third, at which point the state acquires a "compelling" interest in the protection of new life. The Court's inclination to permit abortion in the first two trimesters and all but ban it in the third was both morally defensible and helpful to physicians and regulators, because the beginning of the third trimester can be objectively determined within a week or so. Whatever one thinks of the legal reasoning in *Roe*— the opinion is sometimes attacked even by liberal scholars for its shaky use of precedent — its attempts at rights-balancing are a model of conscientious jurisprudence.

The problem is that *Roe*'s third-trimester protections were brushed aside by two descendent Supreme Court cases, *Danforth* in 1976 and *Colautti* in 1979. *Danforth* tossed out *Roe*'s clear, comprehensible third-trimester distinction and substituted a "viability" standard so vague it was impossible to make heads or tails of it. Unlike the third trimester, which can be objectively delineated, viability is subjective. Some babies are viable at the biological frontier of twenty-three weeks; others die even if carried to term; there is no way to know in advance. *Danforth* went so far as to prohibit states from drawing clear lines at the third trimester — that is, it forbade states from using the logic of *Roe.*

Three years later, *Colautti* essentially said that, since no one could understand *Danforth,* it would henceforth be up to each woman's physician to determine whether a fetus was viable and thus legally protected. Here a misjudgment was poised atop an error, given that no doctor can ever be sure that a fetus is viable. Since the person making the determination may also perform the abortion, all a physician has to do under *Colautti* is hazard a guess that the fetus is not viable, and a late-term abortion may proceed. There is no accountability for, or review of, the physician's judgment. And, if an abortion occurs, no disproof of the doctor's judgment is possible, since the chance of viability ends.

The blurry viability standard was reinforced in the 1992 Supreme Court case called *Casey.* Again the Court appeared to outlaw late-term abortion, saying that a viable fetus should be constitutionally protected. But it rejected bright-line definitions of the onset of life, specifically forbidding states to employ the third trimester as a clear, enforceable standard. Instead the Court cryptically declared that viability confers protection "whenever it may occur" — medically close to meaningless, since there is no sure means to determine viability. Under *Casey*, as under *Colautti*, it is the abortion provider who deems whether a third-trimester fetus is viable, which makes almost any late-term abortion permissible. That is the status quo today.

Casey appeared to grant states the authority to restrict late-term abortion so long as they did not "unduly" burden women seeking early abortions. Thirty states proceeded to enact third-trimester restrictions, but most of these have been struck down, either for being too vague to enforce or for containing Trojan-horse language meant to erode *Roe* itself.

Recently, some states have opted for legislation intended solely to prevent a form of late-term abortion called D&X, in which delivery is induced, the fetus is partly born, feet first, and then the skull is crushed and the brains vacuumed out. There is no moral distinction between aborting a late-term fetus via D&X and doing the same via the D&E procedure, in which death occurs within the womb: either both are defensible or neither is. Yet D&X is undeniably barbaric. The AMA has recommended that its members not perform this procedure, adding that "there does not appear to be any identified situation" in which it is required for the health of the mother. In many nations, the technique is unthinkable: Fisk, of the Imperial College School of Medicine, notes, "I've never known a respectable physician who has done a D&X."

This fall, the U.S. Court of Appeals for the Seventh Circuit upheld a Wisconsin statute that prohibited D&X abortion but allowed D&E. (The opinion was by my brother Frank, an appellate judge, who had no connection to the writing of this article.) Editorialists declared that, for the first time, a federal court had "banned" late-term abortion, though the decision did nothing of the kind — it simply found that Wisconsin could regulate types of late-term procedures, so long as women retained the rights delineated under *Roe*. Also this fall, the U.S. Court of Appeals for the Eighth Circuit overturned a Nebraska law that restricted D&X but made no provisions for threats to the life of the mother. Faced with conflicting opinions among the appellate circuits, the Supreme Court said last week that it would hear the late-term abortion issue again, setting the stage for the first important abortion ruling of this generation.

Meanwhile, each year since 1995, Congress has enacted legislation to restrict late-term abortion, and each year President Clinton has either vetoed or threatened to veto it. During the sequence of votes and vetoes, each side has gone out of its way to make itself look bad. Pro-life members of Congress have proposed absolute bans that make no provision for protecting the life of the mother, which undermines their claim to revere life. Senator Diane Feinstein of California, in what was surely one of the all-time lows for American liberalism, brought to the Senate floor a bill intended to affirm a woman's right to terminate a healthy, viable late-term fetus. Both sides have opposed a reasonable middle ground. In 1996, for example, Representative Steny Hoyer of Maryland, a liberal Democrat, offered a bill to ban late-term abortions except when necessary to avert "serious adverse health consequences" to the woman. Rather than rally around this compromise, pro-lifers and pro-choicers mutually assailed it.

In 1997, the AMA declared that third-trimester abortions should not be performed "except in cases of serious fetal anomalies incompatible with life," meaning when the fetus appears fated to die anyway. The AMA supports *Roe,* backs public funding of abortions, and favors availability of RU-486; it simply thinks that, once a fetus can draw its own breath, a new life exists and must be protected. The AMA declaration had a strong influence on centrists such as Democratic Representative Tim Roemer of Indiana, who has called the D&X procedure "inches from infanticide," and Senate Minority Leader Tom Daschle, who in 1997 switched from supporting late-term abortion to opposing it.

Daschle offered a bill that would have prohibited third-trimester abortions except to avoid "grievous injury" to the mother and would have required any physician performing a late-term abortion to certify that the fetus was not viable. Under pressure from pro-choice lobbyists, Clinton offered only tepid, pro forma support for the Daschle bill. Pro-life activists rallied against it, asserting that the "grievous injury" clause could justify abortions based upon a woman's mental rather than physical health. Gridlock has prevailed since.

The issue of mental health is an example of how absolutist thinking cripples both pro-life and pro-choice advocacy. Pro-life forces find it repugnant that a woman might be allowed to terminate a pregnancy to preserve her emotional state, yet it is fair to assume that no man will ever understand the mental-health consequences to a woman of unwanted motherhood. Conversely, pro-choice theory concerns itself only with a woman's mental health during pregnancy, not afterward. A woman who carries an unwanted child to term and then offers the baby for adoption may suffer physical and psychological hardship and social opprobrium — but, for the rest of her life, her conscience will be clear. Pro-choice absolutism takes no account of the mental health of the woman who aborts a viable child and then suffers remorse for an act she cannot undo.

If women's health and freedom represent the blind spot of the pro-life side, the moral standing of the third-trimester fetus — the baby, by that point — is the blind spot of pro-choicers. Pro-choice adherents cite the slippery slope, but that apprehension is an artifact of lobbying and fund-raising, not of law. Clinton, reflecting the absolutist line, has said that late-term abortion is "a procedure that appears inhumane" but that restrictions "would be even more inhumane" because they would lead to the overturning of *Roe.* For those who know what's actually in *Roe* — a trimester system whose very purpose is to allow early choice while protecting late-term babies — this claim is more than a little ironic.

Women are right to fear that political factions are working to efface their rights. Late-term abortion is simply not the ground on which to stage the defense — because, unless the mother's life is at stake, late-term abortion is wrong.

It is time to admit what everyone knows and what the new science makes clear: that third-trimester abortion should be very tightly restricted. The hopelessly confusing viability standard should be dropped in favor of a bright line drawn at the start of the third trimester, when complex fetal brain activity begins. Restricting abortion after that point would not undermine the rights granted by *Roe,* because there is no complex brain activity before the third trimester and thus no slippery slope to start down. Scientifically based late-term abortion restrictions would not enter into law poignant but unprovable spiritual assumptions about the spark of life but would simply protect lives whose humanity is now known.

To be sure, restrictions on late-term abortion would harm the rights of American women, but the harm would be small, while the moral foundation of abortion choice overall would be strengthened by removing the taint of late-term abortion. By contrast, restrictions on early abortions would cause tremendous damage to women's freedom while offering only a hazy benefit to the next generation, since so many pregnancies end naturally anyway. There are costs to either trade-off, but they are costs that a decent society can bear.

Western Europe is instructive in this regard. In most European Union nations, early abortion is not only legal but far less politically contentious than it is here. Yet, in those same countries, late-term abortion is considered infanticide. All European Union nations except France and the United Kingdom ban abortion in the third trimester, except to save the mother's life. And, even where allowed, late-term abortion occurs at one-third the U.S. rate. Western European countries have avoided casting abortion as a duel between irresolvable absolutes. They treat abortion in the first two trimesters as a morally ambiguous private matter, while viewing it in the third trimester as public and morally odious. We should follow their lead. All it requires is knowledge of the new fetal science and a return to the true logic of *Roe.*

MALCOLM GLADWELL

Baby Steps

FROM *The New Yorker*

IN APRIL OF 1997, Hillary Clinton was the host of a day-long conference at the White House entitled "What New Research on the Brain Tells Us About Our Youngest Children." In her opening remarks, which were beamed live by satellite to nearly a hundred hospitals, universities, and schools, in thirty-seven states, Mrs. Clinton said, "Fifteen years ago, we thought that a baby's brain structure was virtually complete at birth." She went on:

> Now we understand that it is a work in progress, and that everything we do with a child has some kind of potential physical influence on that rapidly forming brain. A child's earliest experiences — their relationships with parents and caregivers, the sights and sounds and smells and feelings they encounter, the challenges they meet — determine how their brains are wired. . . . These experiences can determine whether children will grow up to be peaceful or violent citizens, focused or undisciplined workers, attentive or detached parents themselves.

At the afternoon session of the conference, the keynote speech was given by the director-turned-children's-advocate Rob Reiner. His goal, Reiner told the assembled, was to get the public to "look through the prism" of the first three years of life "in terms of problem solving at every level of society":

> If we want to have a real significant impact, not only on children's success in school and later on in life, healthy relationships, but also an impact on reduction in crime, teen pregnancy, drug abuse, child abuse, welfare, homelessness, and a variety of other social ills, we are going to

have to address the first three years of life. There is no getting around it. All roads lead to Rome.

The message of the conference was at once hopeful and a little alarming. On the one hand, it suggested that the right kind of parenting during those first three years could have a lasting effect on a child's life; on the other hand, it implied that if we missed this opportunity the resulting damage might well be permanent. Today, there is a zero-to-three movement, made up of advocacy groups and policymakers like Hillary Clinton, which uses the promise and the threat of this new brain science to push for better pediatric care, early childhood education, and daycare. Reiner has started something called the I Am Your Child Foundation, devoted to this cause, and has enlisted the support of, among others, Tom Hanks, Robin Williams, Billy Crystal, Charlton Heston, and Rosie O'Donnell. Some lawmakers now wonder whether programs like Head Start ought to be drastically retooled, to focus on babies and toddlers rather than on preschoolers. The state of California recently approved a fifty-cent-per-pack tax on cigarettes to fund programs aimed at improving care for babies and toddlers up to the age of five. The state governments of Georgia and Tennessee send classical music CDs home from the hospital with every baby, and Florida requires that daycare centers play classical music every day — all in the belief that Mozart will help babies build their minds in this critical window of development. "During the first part of the twentieth century, science built a strong foundation for the physical health of our children," Mrs. Clinton said in her speech that morning. "The last years of this century are yielding similar breakthroughs for the brain. We are . . . coming closer to the day when we should be able to insure the well-being of children in every domain — physical, social, intellectual, and emotional."

The first lady took pains not to make the day's message sound too extreme. "I hope that this does not create the impression that, once a child's third birthday rolls around, the important work is over," she said, adding that much of the brain's emotional wiring isn't completed until adolescence, and that children never stop needing the love and care of their parents. Still, there was something odd about the proceedings. This was supposed to be a meeting devoted to new findings in brain science, but hardly any of the brain science that was discussed was new. In fact, only a modest amount of brain science was discussed at all. Many of the speakers were from the worlds of education and policy. Then, there was Mrs. Clinton's claim that the experiences of our first few years could "determine" whether we grow up to be peaceful or violent, focused or undisciplined. We tend to think that the environmental influences upon the way we turn out are the sum of a lifetime of experiences — that someone is disciplined because he spent four years in the Marines, or because he got up every morning as a teenager to train with the swim team. But Hillary Clinton was proposing that we direct our attention instead to what happens to children in a very brief window early in life. The first lady, now a candidate for the United States Senate, is associating herself with a curious theory of human development. Where did this idea come from? And is it true?

John Bruer tackles both these questions in his new book, *The Myth of the First Three Years* (Free Press). From its title, Bruer's work sounds like a rant. It isn't. Noting the cultural clout of the zero-to-three idea, Bruer, who heads a medical research foundation in St. Louis, sets out to compare what people like Rob Reiner and Hillary Clinton are saying to what neuroscientists have actually concluded. The result is a superb book, clear and engaging, that serves as both popular science and intellectual history.

Mrs. Clinton and her allies, Bruer writes, are correct in their premise: the brain at birth *is* a work in progress. Relatively few connections among its billions of cells have yet been established. In the first few years of life, the brain begins to wire itself up at a furious pace, forming hundreds of thousands, even millions, of new synapses every second. Infants produce so many new neural connections, so quickly, that the brain of a two-year-old is actually far more dense with neural connections than the brain of an adult. After three, that burst of activity seems to slow down, and our brain begins the long task of rationalizing its communications network, finding those connections that seem to be the most important and getting rid of the rest.

During this brief initial period of synaptical "exuberance," the brain is especially sensitive to its environment. David Hubel and Torsten Wiesel, in a famous experiment, sewed one of the eyes of a kitten shut for the first three months of its life, and when they opened it back up they found that the animal was permanently blind in that eye. There are critical periods early in life, then, when the brain will not develop properly unless it receives a certain amount of outside stimulation. In another series of experiments, begun in the early seventies, William Greenough, a psychologist at the University of Illinois, showed that a rat reared in a large, toy-filled cage with other rats ended up with a substantially more developed visual cortex than a rat that spent its first month alone in a small, barren cage: the brain, to use the word favored by neuroscientists, is *plastic*— that is, modifiable by experience. In other words, Hillary Clinton's violent citizens and unfocused workers might seem to be the human equivalents of kittens who've had an eye sewed shut, or rats who've been reared in a barren cage. If in the critical first three years of synapse formation we could give people the equivalent of a big cage full of toys, she was saying, we could make them healthier and smarter.

Put this way, these ideas sound quite reasonable, and it's easy to see why they have attracted such excitement. But Bruer's contribution is to show how, on several critical points, this account of child development exaggerates or misinterprets the available evidence.

Consider, he says, the matter of synapse formation. The zero-to-three activists are convinced that the number of synapses we form in our earliest years plays a big role in determining our mental capacity. But do we know that to be true? People with a form of mental retardation known as fragile-X syndrome, Bruer notes, have higher numbers of synapses in their brain than the rest of us. More important, the period in which humans gain real intellectual maturity is late adolescence, by which time the brain is aggressively *pruning* the number of connections. Is intelligence associated with how many synapses you have or with how efficiently you manage to sort out and make sense of those connections later in life? Nor do we know how dependent the initial burst of synapse formation is on environmental stimulation. Bruer writes of an experiment where the right hand of a monkey was restrained in a leather mitten from birth to four months, effectively limiting all sensory stimulation. That's the same period when young monkeys form enormous numbers of connections in the somatosensory cortex, the area of the monkey brain responsible for size and texture discriminations, so you'd think that the restrained hand would be impaired. But it wasn't: within a short time, it was functioning normally, which suggests that there is a lot more flexibility and resilience in some aspects of brain development than we might have imagined.

Bruer also takes up the question of early childhood as a developmental window. It makes sense that if children don't hear language by the age of eleven or twelve they aren't going to speak, and that children who are seriously neglected throughout their upbringing will suffer permanent emotional injury. But why, Bruer asks, did advocates arrive at three years of age as a cutoff point? Different parts of the brain develop at different speeds. The rate of synapse formation in our visual cortex peaks at around three or four months. The synapses in our prefrontal cortex — the parts of our brain involved in the most sophisticated cognitive tasks — peak perhaps as late as three years, and aren't pruned back until middle-to-late adolescence. How can the same cutoff apply to both regions?

Greenough's rat experiments are used to support the critical-window idea, because he showed that he could affect brain development in those early years by altering the environment of his animals. The implications of the experiment aren't so straightforward, though. The experiments began when the rats were about three weeks old, which is already past rat "infancy," and continued until they were fifty-five days old, which put them past puberty. So the experiment showed the neurological consequences of deprivation not during some critical window of infancy but during the creature's entire period of maturation. In fact, when Greenough repeated his experiment with rats that were 450 days old — well past middle age — he found that those kept in complex environments once again had significantly denser neural connections than those kept in isolation.

Even the meaning of the kitten with its eye sewn shut turns out to be far from obvious. When that work was repeated on monkeys, researchers found that if they deprived *both* eyes of early stimulation — rearing a monkey in darkness for its first six months — the animal could see (although not perfectly), and the binocularity of its vision, the ability of its left and right eyes to coordinate images, was normal. The experiment doesn't show that more stimulation is better than less for binocular vision. It just suggests that whatever stimulation there is should be balanced, which is why closing one eye tilts the developmental process in favor of the open eye.

To say that the brain is plastic, then, is not to say that the brain is dependent on certain narrow windows of stimulation. Neuroscientists say instead that infant brains have "experience-expectant plasticity" — which means that they need only something that approximates a normal environment. Bruer writes:

> The odds that our children will end up with appropriately fine-tuned brains are incredibly favorable, because the stimuli the brain *expects* during critical periods are the kinds of stimuli that occur everywhere and all the time within the normal developmental environment for our species. It is only when there are severe genetic or environmental aberrations from the normal that nature's expectations are frustrated and neural development goes awry.

In the case of monkeys, the only way to destroy their binocular vision is to sew one eye shut for six months — an entirely contrived act that would almost never happen in the wild. Greenough points out that the "complex" environment he created for his rats — a large cage full of toys and other animals — is actually the closest equivalent of the environment that a rat would encounter naturally. When he created a super-enriched environment for his rats, one with even more stimulation than they would normally encounter, the rats weren't any better off. The only way he could affect the neurological development of the animals was to put them in a barren cage by themselves — again, a situation that an animal would never encounter in the wild. Bruer quotes Steve Petersen, a neuroscientist at Washington University, in St. Louis, as saying that neurological development so badly *wants* to happen that his only advice to parents would be "Don't raise your children in a closet, starve them, or hit them in the head with a frying pan." Petersen was, of course, being flip. But the general conclusion of researchers seems to be that we human beings enjoy a fairly significant margin of error in our first few years of life. Studies done of Romanian orphans who spent their first year under conditions of severe deprivation suggest that most (but not all) can recover if adopted into a nurturing home. In another study, psychologists examined children from an overcrowded orphanage who had been badly neglected as infants and subsequently adopted into loving homes. Within two years of their adoption, one psychologist involved in their rehabilitation had concluded:

> We had not anticipated the older children who had suffered deprivations for periods of two and a half to four years to show swift response to treatment. That they did so amazed us. These inarticulate, underdeveloped youngsters who had formed no relationships in their lives, who were aimless and without a capacity to concentrate on anything, had resembled a pack of animals more than a group of human beings. . . . As we worked with the children, it became apparent that their inadequacy was not the result of damage but, rather, was due to a dearth of normal experiences without which development of human qualities is impossible. After a year of treatment, many of these older children were showing a trusting dependency toward the staff of volunteers and . . . self-reliance in play and routines.

Some years ago, the Berkeley psychology professor Alison Gopnik and one of her students, Betty Repacholi, conducted an experiment with a series of fourteen-month-old toddlers. Repacholi showed the babies two bowls of food, one filled with Goldfish crackers and one filled with raw broccoli. All the babies, naturally, preferred the crackers. Repacholi then tasted the two foods, saying "Yuck" and making a disgusted face at one and saying "Yum" and making a delighted face at the other. Then she pushed both bowls toward the babies, stretched out her hand, and said, "Could you give me some?"

When she liked the crackers, the babies gave her crackers. No surprise there. But when Repacholi liked the broccoli and hated the crackers, the babies were presented with a difficult philosophical issue — that different people may have different, even conflicting, desires. The fourteen-month-olds couldn't grasp that. They thought that if they liked crackers everyone liked crackers, and so they gave Repacholi the crackers, despite her expressed preferences. Four months later, the babies had, by and large, figured this principle out, and when Repacholi made a face at the crackers they knew enough to give her the broccoli. *The Scientist in the Crib* (Morrow), a fascinating new book that Gopnik has written with Patricia Kuhl and Andrew Meltzoff, both at the University of Washington, argues that the discovery of this principle — that different people have different desires — is the source of the so-called terrible twos. "What makes the terrible twos so terrible is not that the babies do things you don't want them to do — one-year-olds are plenty good at that — but that they do things *because* you don't want them to," the authors write. And why is that? Not, as is commonly thought, because toddlers want to test parental authority, or because they're just contrary. Instead, the book argues, the terrible twos represent a rational and engaged exploration of what is to two-year-olds a brand-new idea — a generalization of the insight that the fact that they hate broccoli and like crackers doesn't mean that everyone hates broccoli and likes crackers. "Toddlers are systematically testing the dimensions on which their desires and the desires of others may be in conflict," the authors write. Infancy is an experimental research program, in which "the child is the budding psychologist; we parents are the laboratory rats."

These ideas about child development are, when you think about it, oddly complementary to the neurological arguments of John Bruer. The paradox of the zero-to-three movement is that, for all its emphasis on how alive children's brains are during their early years, it views babies as profoundly passive — as hostage to the quality of the experiences provided for them by their parents and caregivers. *The Scientist in the Crib* shows us something quite different. Children are scientists, who develop theories and interpret evidence from the world around them in accordance with those theories. And when evidence starts to mount suggesting that the existing theory isn't correct — wait a minute, just because I like crackers doesn't mean Mommy likes crackers — they create a new theory to explain the world, just as a physicist would if confronted with new evidence on the relation of energy and matter. Gopnik, Meltzoff, and Kuhl play with this idea at some length. Science, they suggest, is actually a kind of institutionalized childhood, an attempt to harness abilities that evolved to be used by babies or young children. Ultimately, the argument suggests that child development is a rational process directed and propelled by the child himself. How does the child learn about different desires? By systematically and repeatedly provoking a response from adults. In the broccoli experiment, the adult provided the fourteen-month-old with the information ("I hate Goldfish crackers") necessary to make the right decision. But the child ignored that information until he himself had developed a theory to interpret it. When *The Scientist in the Crib* describes children as budding psychologists and adults as laboratory rats, it's more than a clever turn of phrase. Gopnik, Meltzoff, and Kuhl observe that our influence on infants "seems to work in concert with children's own learning abilities." Newborns will "imitate facial expressions" but not "complex actions they don't understand themselves." And the authors conclude, "Children won't take in what you tell them until it makes sense to them. Other people don't simply shape what children do; parents aren't the programmers. Instead, they seem designed to provide just the right sort of information."

It isn't until you read *The Scientist in the Crib* alongside more conventional child-development books that you begin to appreciate the full implications of its argument. Here, for example, is a passage from *What's Going On in There? How the Brain and Mind Develop in the First Five Years of Life,* by Lise Eliot, who teaches at the University of Chicago: "It's important to avoid the kind of muddled baby-talk that turns a sentence like 'Is she the cutest little baby in the world?' into 'Uz see da cooest wiwo baby inna wowud?' Caregivers should try to enunciate clearly when speaking to babies and young children, giving them the cleanest, simplest model of speech possible." Gopnik, Meltzoff, and Kuhl see things a little differently. First, they point out, by six or seven months babies are already highly adept at decoding the sounds they hear around them, using the same skills we do when we talk to someone with a thick foreign accent or a bad cold. If you say "Uz see da cooest wiwo baby inna wowud?" they hear something like "Is she the cutest little baby in the world?" Perhaps more important, this sort of Motherese — with its elongated vowels and repetitions and overpronounced syllables — is just the thing for babies to develop their language skills. And Motherese, the authors point out, seems to be innate. It's found in every culture in the world, and anyone who speaks to a baby uses it, automatically, even without realizing it. Babies want Motherese, so they manage to elicit it from the rest of us. That's a long way from the passive baby who thrives only because of the specialized, high-end parenting skills of the caregiver. "One thing that science tells us is that nature has designed us to teach babies, as much as it has designed babies to learn," Gopnik, Meltzoff, and Kuhl write. "Almost all of the adult actions we've described" — actions that are critical for the cognitive development of babies — "are swift, spontaneous, automatic and unpremeditated."

Does it matter that Mrs. Clinton and her allies have misread the evidence on child development? In one sense, it doesn't. The first lady does not claim to be a neuroscientist. She is a politician, and she is interested in the brains of children only to further an entirely worthy agenda: improved daycare, pediatric care, and early-childhood education. Sooner or later, however, bad justifications for social policy can start to make for bad social policy, and that is the real danger of the zero-to-three movement.

In Lise Eliot's book, for instance, there's a short passage in which she writes of the extraordinary powers of imitation that infants possess. A fifteen-month-old who watches an experimenter lean over and touch his forehead to the top of a box will, when presented with that same box four months later, do exactly the same thing. "The fact that these memories last so long is truly remarkable — and a little bit frightening," Eliot writes, and she continues:

> It goes a long way toward explaining why children, even decades later, are so prone to replicating their parents' behavior. If toddlers can repeat, even several months later, actions they've seen only once or twice, just imagine how watching their parents' daily activities must affect them. Everything they see and hear over time — work, play, fighting, smoking, drinking, reading, hitting, laughing, words, phrases, and gestures — is stored in ways that shape their later actions, and the more they see of a particular behavior, the likelier it is to reappear in their own conduct.

There is something to this. Why we act the way we do is obviously the result of all kinds of influences and experiences, including those cues we pick up unconsciously as babies. But this doesn't mean, as Eliot seems to think it does, that you can draw a straight line between a concrete adult behavior and what little Suzie, at six months, saw her mother do. As far as we can tell, for instance, infant imitation has nothing to do with smoking. As the behavioral geneticist David Rowe has demonstrated, the children of smokers are more likely than others to take up the habit because of genetics: they have inherited the same genes that made their parents like, and be easily addicted to, nicotine. Once you account for heredity, there is little evidence that parental smoking habits influence children; the adopted children of smokers, for instance, are no more likely to smoke than the children of nonsmokers. To the extent that social imitation is a factor in smoking, the psychologist Judith Rich Harris has observed, it is imitation that occurs in adolescence between a teenager and his or her peers. So if you were to use Eliot's ideas to design an antismoking campaign you'd direct your efforts to stop parents from smoking around their children, and miss the social roots of smoking entirely.

This point — the distance between infant experience and grownup behavior — is made even more powerfully in Jerome Kagan's marvelous new book, *Three Seductive Ideas* (Oxford). Kagan, a professor of psychology at Harvard, offers a devastating critique of what he calls "infant determinism," arguing that many of the truly critical moments of socialization — the moments that social policy properly concerns itself with — occur well after the age of three. As Kagan puts it, a person's level of "anxiety, depression, apathy and anger" is linked to his or her "symbolic constructions of experience" — how the bare facts of any experience are combined with the context of that event, attitudes toward those involved, expectations, and memories of past experience. "The Palestinian youths who throw stones at Israeli soldiers believe that the Israeli government has oppressed them unjustly," Kagan writes. He goes on:

> The causes of their violent actions are not traceable to the parental treatment they received in their first few years. Similarly, no happy African-American two-year-old knows about the pockets of racism in American society or the history of oppression blacks have suffered. The realization that there is prejudice will not take form until that child is five or six years old.

Infant determinism doesn't just encourage the wrong kind of policy. Ultimately, it undermines the basis of social policy. Why bother spending money trying to help older children or adults if the patterns of a lifetime are already, irremediably, in place? Inevitably, some people will interpret the zero-to-three dogma to mean that our obligations to the disadvantaged expire by the time they reach the age of three. Kagan writes of a famous Hawaiian study of child development, in which almost 700 children, from a variety of ethnic and economic backgrounds, were followed from birth to adulthood. The best predictor of who would develop serious academic or behavioral problems in adolescence, he writes, was social class: more than 80 percent of the children who got in trouble came from the poorest segment of the sample. This is the harsh reality of child development, from which the zero-to-three movement offers a convenient escape. Kagan writes, "It is considerably more expensive to improve the quality of housing, education and health of the approximately one million children living in poverty in America today than to urge their mothers to kiss, talk to, and play with them more consistently." In his view, "to suggest to poor parents that playing with and talking to their infant will protect the child from future academic failure and guarantee life success" is an act of dishonesty. But that does not go far enough. It is also an unwitting act of reproach: it implies to disadvantaged parents that if their children do not turn out the way children of privilege do it is their fault — that they are likely to blame for the flawed wiring of their children's brains.

In 1973, when Hillary Clinton — then, of course, known as Hillary Rodham — was a young woman just out of law school, she wrote an essay for the *Harvard Educational Review* entitled "Children Under the Law." The courts, she wrote, ought to reverse their longstanding presumption that children are legally incompetent. She urged, instead, that children's interests be considered independently from those of their parents. Children ought to be deemed capable of making their own decisions and voicing their own interests, unless evidence could be found to the contrary. To her, the presumption of incompetence gave the courts too much discretion in deciding what was in the child's best interests, and that discretion was most often abused in cases of children from poor minority families. "Children of these families," she wrote, "are perceived as bearers of the sins and disabilities of their fathers."

This is a liberal argument, because a central tenet of liberalism is that social mobility requires a release not merely from burdens imposed by poverty but also from those imposed by family — that absent or indifferent or incompetent parents should not be permitted to destroy a child's prospects. What else was the classic Horatio Alger story about? In *Ragged Dick,* the most famous of Alger's novels, Dick's father runs off before his son's birth, and his mother dies destitute while Dick is still a baby. He becomes a street urchin, before rising to the middle class through a combination of hard work, honesty, and luck. What made such tales so powerful was, in part, the hopeful notion that the circumstances of your birth need not be your destiny; and the modern liberal state has been an attempt to make good on that promise.

But Mrs. Clinton is now promoting a movement with a different message — that who you are and what you are capable of could be the result of how successful your mother and father were in rearing you. In her book *It Takes a Village,* she criticizes the harsh genetic determinism of *The Bell Curve.* But an ideology that holds that your future is largely decided at birth by your parents' genes is no more dispiriting than one that holds that your future might be decided at three by your parents' behavior. The unintended consequence of the zero-to-three movement is that, once again, it makes disadvantaged children the bearers of the sins and disabilities of their parents.

The truth is that the traditional aims of the liberal agenda find ample support in the arguments of John Bruer, of Jerome Kagan, of Judith Rich Harris, and of Gopnik, Meltzoff, and Kuhl. All of them offer considerable evidence that what the middle class perceives as inadequate parenting need not condemn a baby for life, and that institutions and interventions to help children as they approach maturity can make a big difference in how they turn out. It is, surely, a sad irony that, at the very moment when science has provided the intellectual reinforcement for modern liberalism, liberals themselves are giving up the fight.

JANE GOODALL

In the Forests of Gombe

FROM *Orion*

I WAS TAUGHT, AS A SCIENTIST, to think logically and empirically, rather than intuitively or spiritually. When I was at Cambridge University in the early 1960s most of the scientists and science students working in the Department of Zoology, so far as I could tell, were agnostic or even atheist. Those who believed in a god kept it hidden from their peers.

Fortunately, by the time I got to Cambridge I was twenty-seven years old and my beliefs had already been molded so that I was not influenced by these opinions. I believed in the spiritual power that, as a Christian, I called God. But as I grew older and learned about different faiths I came to believe that there was, after all, but One God with different names: Allah, Tao, the Creator, and so on. God, for me, was the Great Spirit in Whom "we live and move and have our being." There have been times during my life when this belief wavered, when I questioned — even denied — the existence of God. At such times I felt there can be no underlying meaning to the emergence of life on earth.

Still, for me those periods have been relatively rare, triggered by a variety of circumstances. One was when my second husband died of cancer. I was grieving, suffering, and angry. Angry at God, at fate — the unjustness of it all. For a time I rejected God, and the world seemed a bleak place.

It was in the forests of Gombe that I sought healing after Derek's death. Gradually during my visits, my bruised and battered spirit found solace. In the forest, death is not hidden — or only accidentally, by the fallen leaves. It is all around you all the time, a part of the endless cycle of life. Chimpanzees are born, they grow older, they get sick, and they die. And always there are the young ones to carry on the life of the species. Time spent in the forest, following and watching and simply being with the chimpanzees, has always sustained the inner core of my being. And it did not fail me then.

One day, among all the days, I remember most of all. It was May 1981 and I had finally made it to Gombe after a six-week tour in America — six weeks of fund-raising dinners, conferences, meetings, and lobbying for various chimpanzee issues. I was exhausted and longed for the peace of the forest. I wanted nothing more than to be with the chimpanzees, renewing my acquaintance with my old friends, getting my climbing legs back again, relishing the sights, sounds, and smells of the forest. I was glad to be away from Dar es Salaam, with all its sad associations — the house that Derek and I had shared, the palm trees we had bought and planted together, the rooms we had lived in together, the Indian Ocean in which Derek, handicapped on land, had found freedom swimming among his beloved coral reefs.

Back in Gombe. It was early in the morning and I sat on the steps of my house by the lakeshore. It was very still. Suspended over the horizon, where the mountains of the Congo fringed Lake Tanganyika, was the last quarter of the waning moon and her path danced and sparkled toward me across the gently moving water. After enjoying a banana and a cup of coffee, I was off, climbing up the steep slopes behind my house.

In the faint light from the moon reflected by the dew-laden grass, it was not difficult to find my way up the mountain. It was quiet, utterly peaceful. Five minutes later I heard the rustlings of leaves overhead. I looked up and saw the branches moving against the lightening sky. The chimps had awakened. It was Fifi and her offspring, Freud, Frodo, and little Fanni. I followed when they moved off up the slope, Fanni riding on her mother's back like a diminutive jockey. Presently they climbed into a tall fig tree and began to feed. I heard the occasional soft thuds as skins and seeds of figs fell to the ground.

For several hours we moved leisurely from one food tree to the next, gradually climbing higher and higher. On an open grassy ridge the chimps climbed into a massive mbula tree, where Fifi, replete from the morning's feasting, made a large comfortable nest high above me. She dozed through a midday siesta, little Fanni asleep in her arms, Frodo and Freud playing nearby. I felt very much in tune with the chimpanzees, for I was spending time with them not to observe, but simply because I needed their company, undemanding and free of pity. From where I sat I could look out over the Kasakela Valley. Just below me to the west was the peak. From that same vantage point I had learned so much in the early days, sitting and watching while, gradually, the chimpanzees had lost their fear of the strange white ape who had invaded their world. I recaptured some of my long-ago feelings — the excitement of discovering, of seeing things unknown to Western eyes, and the serenity that had come from living, day after day, as a part of the natural world. A world that dwarfs yet somehow enhances human emotions.

As I reflected on these things I had been only partly conscious of the approach of a storm. Suddenly, I realized that it was no longer growling in the distance but was right above. The sky was dark, almost black, and the rain clouds had obliterated the higher peaks. With the growing darkness came the stillness, the hush, that so often precedes a tropical downpour. Only the rumbling of the thunder, moving closer and closer, broke this stillness; the thunder and the rustling movements of the chimpanzees. All at once came a blinding flash of lightning, followed, a split second later, by an incredibly loud clap of thunder that seemed almost to shake the solid rock before it rumbled on, bouncing from peak to peak. Then the dark and heavy clouds let loose such torrential rain that sky and earth seemed joined by moving water. I sat under a palm whose fronds, for a while, provided some shelter. Fifi sat hunched over, protecting her infant; Frodo pressed close against them in the nest; Freud sat with rounded back on a nearby branch. As the rain poured endlessly down, my palm fronds no longer provided shelter and I got wetter and wetter. I began to feel first chilly, and then, as a cold wind sprang up, freezing; soon, turned in on myself, I lost all track of time. I and the chimpanzees formed a unit of silent, patient, and uncomplaining endurance.

It must have been an hour or more before the rain began to ease as the heart of the storm swept away to the south. At four-thirty the chimps climbed down, and we moved off through the dripping vegetation, back down the mountainside. Presently we arrived on a grassy ridge overlooking the lake. I heard sounds of greeting as Fifi and her family joined Melissa and hers. They all climbed into a low tree to feed on fresh young leaves. I moved to a place where I could stand and watch as they enjoyed their last meal of the day. Down below, the lake was still dark and angry with white flecks where the waves broke, and rain clouds remained black in the south. To the north the sky was clear with only wisps of gray clouds still lingering. In the soft sunlight, the chimpanzees' black coats were shot with coppery brown, the branches on which they sat were wet and dark as ebony, the young leaves a pale but brilliant green. And behind was the backcloth of the indigo sky where lightning flickered and distant thunder growled and rumbled.

Lost in awe at the beauty around me, I must have slipped into a state of heightened awareness. It is hard — impossible, really — to put into words the moment of truth that suddenly came upon me then. It seemed to me, as I struggled afterward to recall the experience, that *self* was utterly absent: I and the chimpanzees, the earth and trees and air, seemed to merge, to become one with the spirit power of life itself. The air was filled with a feathered symphony, the evensong of birds. I heard new frequencies in their music and also in the singing insects' voices — notes so high and sweet I was amazed. Never had I been so intensely aware of the shape, the color of the individual leaves, the varied patterns of the veins that made each one unique. Scents were clear as well, easily identifiable: fermenting overripe fruit; waterlogged earth; cold, wet bark; the damp odor of chimpanzee hair and, yes, my own too. I sensed a new presence, then saw a bushbuck, quietly browsing upwind, his spiraled horns gleaming and chestnut coat dark with rain.

Suddenly a distant chorus of pant-hoots elicited a reply from Fifi. As though wakening from some vivid dream I was back in the everyday world, cold, yet intensely alive. When the chimpanzees left, I stayed in that place — it seemed a most sacred place — scribbling some notes, trying to describe what, so briefly, I had experienced.

Eventually I wandered back along the forest trail and scrambled down behind my house to the beach. Later, as I sat by my little fire, cooking my dinner of beans, tomatoes, and an egg, I was still lost in the wonder of my experience. Yes, I thought, there are many windows through which we humans, searching for meaning, can look out into the world around us. There are those carved out by Western science, their panes polished by a succession of brilliant minds. Through them we can see ever farther, ever more clearly, into areas which until recently were beyond human knowledge. Through such a scientific window I had been taught to observe the chimpanzees. For more than twenty-five years I had sought, through careful recording and critical analysis, to piece together their complex social behavior, to understand the workings of their minds. And this had not only helped us to better understand their place in nature but also helped us to understand a little better some aspects of our own human behavior, our own place in the natural world.

Yet there are other windows through which we humans can look out into the world around us, windows through which the mystics and the holy men of the East, and the founders of the great world religions, have gazed as they searched for the meaning and purpose of our life on earth, not only in the wondrous beauty of the world, but also in its darkness and ugliness. And those Masters contemplated the truths that they saw, not with their minds only but with their hearts and souls also. From those revelations came the spiritual essence of the great scriptures, the holy books, and the most beautiful mystic poems and writings. That afternoon it had been as though an unseen hand had drawn back a curtain and, for the briefest moment, I had seen through such a window.

How sad that so many people seem to think that science and religion are mutually exclusive. Science has used modern technology and modern techniques to uncover so much about the formation and the development of life forms on Planet Earth and about the solar system of which our little world is but a minute part. Alas, all of these amazing discoveries have led to a belief that every wonder of the natural world and of the universe — indeed, of infinity and time — can, in the end, be understood through the logic and the reasoning of a finite mind. And so, for many, science has taken the place of religion. It was not some intangible God who created the universe, they argue, it was the big bang. Physics, chemistry, and evolutionary biology can explain the start of the universe and the appearance and progress of life on earth, they say. To believe in God, in the human soul, and in life after death is simply a desperate and foolish attempt to give meaning to our lives.

But not all scientists believe thus. There are quantum physicists who have concluded that the concept of God is not, after all, merely wishful thinking. There are those exploring the human brain who feel that no matter how much they discover about this extraordinary structure it will never add up to a complete understanding of the human mind — that the whole is, after all, greater than the sum of its parts. The big bang theory is yet another example of the incredible, the awe-inspiring ability of the human mind to learn about seemingly unknowable phenomena in the beginning of time. Time as we know it, or think we know it. But what about before time? And what about beyond space? I remember so well how those questions had driven me to distraction when I was a child.

I lay flat on my back and looked up into the darkening sky. I thought about the young man I had met during the six-week tour I had finished before my return to Gombe. He had a holiday job working as a bellhop in the big hotel where I was staying in Dallas, Texas. It was prom night, and I wandered down to watch the young girls in their beautiful evening gowns, their escorts elegant in their tuxedos. As I stood there, thinking about the future — theirs, mine, the world's — I heard a diffident voice:

"Excuse me, Doctor — aren't you Jane Goodall?" The bellhop was very young, very fresh-faced. But he looked worried — partly because he felt that he should not be disturbing me, but partly, it transpired, because his mind was indeed troubled. He had a question to ask me. So we went and sat on some back stairs, away from the glittering groups and hand-holding couples.

He had watched all my documentaries, read my books. He was fascinated, and he thought that what I did was great. But I talked about evolution. Did I believe in God? If so, how did that square with evolution? Had we really descended from chimpanzees?

And so I tried to answer him as truthfully as I could, to explain my own beliefs. I told him that no one thought humans had descended from chimpanzees. I explained that I did believe in Darwinian evolution and told him of my time at Olduvai, when I had held the remains of extinct creatures in my hands. That I had traced, in the museum, the various stages of the evolution of, say, a horse: from a rabbit-sized creature that gradually, over thousands of years, changed, became better and better adapted to its environment, and eventually was transformed into the modern horse. I told him I believed that millions of years ago there had been a primitive, apelike, humanlike creature, one branch of which had gone on to become the chimpanzee, another branch of which had eventually led to us.

"But that doesn't mean I don't believe in God," I said. And I told him something of my beliefs, and those of my family. I told him that I had always thought that the biblical description of God creating the world in seven days might well have been an attempt to explain evolution in a parable. In that case, each of the days would have been several million years.

"And then, perhaps, God saw that a living being had evolved that was suitable for His purpose. *Homo sapiens* had the brain, the mind, the potential. Perhaps," I said, "that was when God breathed the Spirit into the first Man and the first Woman and filled them with the Holy Ghost."

The bellhop was looking considerably less worried. "Yes, I see," he said. "That could be right. That does seem to make sense."

I ended by telling him that it honestly didn't matter how we humans got to be the way we are, whether evolution or special creation was responsible. What mattered and mattered desperately was our future development. How should the mind that can contemplate God relate to our fellow beings, the other life forms of the world? What is our human responsibility? And what, ultimately, is our human destiny? Were we going to go on destroying God's creation, fighting each other, hurting the other creatures of His planet? Or were we going to find ways to live in greater harmony with each other and with the natural world? That, I told him, was what was important. Not only for the future of the human species, but also for him, personally. When we finally parted his eyes were clear and untroubled, and he was smiling.

Thinking about that brief encounter, I smiled too, there on the beach at Gombe. A wind sprang up and it grew chilly. I left the bright stars and went inside to bed. I knew that while I would always grieve Derek's passing, I could cope with my grieving. That afternoon, in a flash of "outsight" I had known timelessness and quiet ecstasy, sensed a truth of which mainstream science is merely a small fraction. And I knew that the revelation would be with me for the rest of my life, imperfectly remembered yet always within. A source of strength on which I could draw when life seemed harsh or cruel or desperate. The forest, and the spiritual power that was so real in it, had given me the "peace that passeth understanding."

JEROME GROOPMAN

The Doubting Disease

FROM *The New Yorker*

ON A SNOWY SUNDAY in winter, I attended a conference in Cambridge, Massachusetts. The participants included a wide variety of scientists: molecular biologists, organic chemists, computer programmers, virologists, clinical researchers, and statisticians. Afterward, a small group of us went to dinner at a local restaurant. During the meal, the conversation turned to schooling.

"I transferred my eight-year-old out of public school last year," a chemist told the group. "The teacher wouldn't accommodate him. My kid is like me. When he has a problem to solve, he attacks it until it's done perfectly. He completely blocks out the world and won't let go. The teacher insisted that he couldn't spend more than the allotted time on a task. When my son wouldn't stop, the teacher concluded that he had a behavior disorder."

This anecdote provoked a startlingly sympathetic response around the table: most of us, it turned out, identified with the chemist's son. A biologist known for deciphering, atom by atom, the three-dimensional structure of complex proteins declared, "I bet I qualify for what psychiatrists call obsessive-compulsive disorder. When I'm reviewing lab data, and especially when I'm ready to send out a scientific paper, I keep thinking something is wrong. I become intensely anxious. I'll stay up all night reworking every graph and equation. I'm unable to get the thought out of my head that there's a mistake. Then I find myself checking other kinds of things. I'll go blocks away from the house and turn back to make sure the doors are locked, even though I know they are." He turned to the chemist. "I'm not sure what would have happened if I had had your son's teacher."

What did it mean, I wondered as I left the restaurant, that a group of prominent scientists showed at least some traits associated with a clinical disorder during periods of high anxiety? More and more American children are being diagnosed and medicated every year, and at younger and younger ages. If my colleagues and I were in school now, would we be considered abnormal?

Current estimates hold that more than 2 percent of the United States population — nearly 7 million people — have or have had obsessive-compulsive disorder (OCD). The American Psychiatric Association classifies all known mental disorders in its *Diagnostic and Statistical Manual,* or *DSM.* Obsessive-compulsive disorder, which usually manifests itself in adolescence, is characterized by recurrent, time-consuming obsessions or compulsions that are severe enough to cause marked distress or significant impairment. Furthermore, the person recognizes that his obsessions or compulsions are excessive or unreasonable. "Obsessions" are defined in the *DSM* as persistent thoughts, impulses, or images that are experienced as intrusive, anxiety-producing, and inappropriate.

A person with such obsessions usually tries to ignore them, or to defuse them with some other thoughts or actions: this attempt defines a compulsion. You're obsessed with the thought that you didn't turn off the stove; you compulsively check to make sure it's off. (The French call OCD "the doubting disease.") Other well-recognized compulsions are hand-washing, counting, or repeating special words. In its extreme form, people afflicted with OCD are virtual prisoners of their compulsions — exhausted, ashamed, alienated from others. Certainly, nobody at the restaurant would have qualified for the diagnosis. Our obsessions tended to be temporary and connected to a productive activity, like solving an equation. We may describe ourselves as "obsessive," but our obsessions don't control us.

Although there is little information about the biological roots of the disorder — some have speculated that it can follow strep infections — recent studies indicate that people with OCD have distinctive neurological circuitry. These differences are most pronounced in the limbic lobe, the caudate nucleus, and the orbital frontal cortex, the areas of the brain that participate in anxiety and automatic responses. Sophisticated brain scans show that when a potentially distressing scenario is confronted by a person without OCD, the brain activity in these areas barely registers on the screen; in a person with OCD, however, there is an intense and prolonged firing of neurons, and the scans light up like a Christmas tree. The Cambridge conference left me wondering whether scientists and other driven, detail-oriented professionals could also have distinctive neurological circuitry. Or are these mildly obsessive-compulsive people more likely to be attracted to these fields?

The next day, I found myself taking another look at the familiar environment of my laboratory. In the lab — where many scientists spend ten to twelve hours each day, six to seven days a week — everything is tightly controlled. Tedious tasks demand absolute concentration, because a single error can wreck months of work. During our lab's weekly meeting, every detail of every experiment is intensely scrutinized and challenged as we search for those hidden, threatening mistakes. Is this the natural habitat of the obsessive-compulsive?

Speaking with a score of fellow scientists throughout the week, I elicited anecdote after anecdote of mildly obsessive-compulsive behavior. One researcher said that when she approaches the lab to prepare for a particularly important experiment, she counts to herself and taps the wall as she walks down the corridor. Another "prefers" prime numbers, and counts to three or to seven before analyzing a sequence of DNA. A third told me that, during the month before her grant proposals are due, she repeatedly returns home to check the stove in her apartment, even though she knows that it is turned off.

I also looked for survey studies on personality traits of scientists, or of children and adolescents who pursue careers in high technology. I searched for published articles in the National Library of Medicine, a repository of clinical literature; I checked listings of hundreds of popular books on Internet booksellers. Nothing specifically addressed the issue. I decided that it was time to seek professional help.

"What is a disorder, anyway?" the psychologist Jane Holmes Bernstein asked me rhetorically, in an animated English accent. Holmes Bernstein is the director of the neuropsychology program at Boston's Children's Hospital, and she is an expert at behavioral assessments of children. Like most scientists, she has a healthy skepticism toward her own field: "I decided early in the game that I needed to be hit with the full battery of neuropsychiatric tests that I give to kids — that it wasn't fair unless I experienced them." One day, when she was testing a child who had been referred to her for certain learning difficulties at school, she realized that he tested exactly as she herself had. "I asked myself, 'Why am I on my side of the desk?' In my environment, I function at a high level, where it plays out adaptively."

Holmes Bernstein argues that personality and behaviors can't be considered separately from the particular worlds in which people live; for that reason, she deemphasizes labels and focuses instead on the relationship between behavior and environment. "Many psychiatrists and psychologists fit kids into diagnostic boxes," she asserted. "This thinking begins in medical school. There is distinctive, intrinsic organic pathology, the patient put into a box labeled 'diabetes' or 'HIV.' But those boxes are not built for behavior, because behavior is influenced so strongly by its interaction with environment."

She suggested that OCD is a response to excess arousal — arousal in this instance meaning a neurological response to environmental stimuli. "The OCD neurological circuits in the limbic system are set higher for certain stimuli and can respond faster," Holmes Bernstein said. She pointed to recent studies at Indiana University that show that, under certain conditions, people with OCD make associations between neutral as well as aversive stimuli more quickly than people without OCD.

Holmes Bernstein believes that both this high state of arousal and the anxiety it produces may have evolutionary roots. In a prehistoric environment, those with the ability to focus and lock onto stimuli — particularly onto threatening elements in the environment — could have been better suited to escape the dangers of predators and treacherous terrain. But only to a point. "An adaptive mechanism can always become nonadaptive," Holmes Bernstein said.

"This argument about the precise definition of OCD is not just semantic, because it is the *DSM* that dictates treatments," she went on. "Left to itself, the human animal accepts a wide range of behavior. OCD becomes as much an issue of managing load in a high-stimulus environment as it is a specific neurological disorder."

*

After leaving Holmes Bernstein, I got in touch with Anthony Rao, a clinical psychologist who has a large community-based practice in child and adolescent psychology in the Boston area. Rao's specialty is behavioral therapy, and he regularly sees children like the chemist's son, who are brought by their parents or referred by teachers. He feels that he is constantly battling against misguided attempts to diagnose children and provide generic remedies. "There is too much pathologizing of people's behaviors," he said. "In the educational system, it's one size fits all. Teachers run to labels, like ADD" — attention-deficit disorder — "or OCD, and even tell parents their children need to start taking medication." Even among preschoolers, as a recent *Journal of the American Medical Association* study showed, there has been a sharp rise in the use of psychiatric medications, not only for ADD but also for putative anxiety and depression.

What drives all this, Rao believes, is the free-floating anxiety that parents — often successful members of the middle and upper class — foist on their children. In the instability of today's global economy, they fear that any deviation from the norm may cripple their child's future. He also believes that the currently fashionable psychiatric model — the idea that the problem is "a chemical disease of the brain" — is overly simplistic and even dangerous. These days, psychiatrists primarily treat OCD with selective serotonin reuptake inhibitors, like Prozac and Luvox, which alleviate not only its symptoms but also the anxiety and depression that often accompany it. But Rao pointed out that no one knows precisely what the long-term effects of these drugs on children will be — "especially when they are given daily for years." This approach, he contends, is treating the brain as if it were a bad kidney, when it's a far more complex organ, one that modifies itself continually.

Rao was careful to stipulate that he does not categorically oppose medication. A child or an adolescent with OCD who can't leave the house, or who can't sleep because he needs to repeatedly check under the bed, may greatly benefit from drugs. The problem is with the larger universe of kids who are summarily labeled "abnormal" and medicated. In March, the White House expressed alarm at this trend, and the National Institute of Mental Health called for new studies to assess the safety and efficacy of psychoactive drug therapy for young children.

"It's a different world in psychiatry now, with managed care," Rao went on. "In order for a psychiatrist to get paid, he needs to give you a *DSM* diagnosis." Rao described a scenario that he often hears from clients about visits to psychiatrists: "Do you have worries? Do you have compulsive acts? Do you realize they are bothersome? Yes, yes, yes. Then it's boom-boom-boom, here's a prescription. You have OCD." Rao believes that the *DSM* label resonates in the child's mind and among family members and friends in pejorative and embarrassing ways. "Your brain is your soul," Rao said fiercely. "You're telling a kid that there is something wrong with who he fundamentally is."

Rao thinks that this excess pathologizing of people's problems is strongly driven by economics: beyond the imperatives of managed care, there is a burgeoning pharmaceutical industry that reaps huge profits from psychotropic medications. Researchers have obvious incentives to conduct drug trials that will encourage the Food and Drug Administration to approve a medication for a specific *DSM*-defined disorder. Rao, on the other hand, attempts to temper his patients' anxieties and to redirect the compulsive behavior into more productive channels.

Rao told me about a recent case of a thirteen-year-old girl whom we'll call Jan, a gifted pianist and a straight-A student. During the past year, her performance at school had plummeted. When she began to exhibit obsessive-compulsive behavior, her parents took her to a psychiatrist, who prescribed OCD medication. The drugs were of little benefit, and friends and teachers began to treat her as though she were seriously disturbed.

The family came to Rao for a second opinion. He learned that she was haunted by thoughts of serious harm coming to her parents. To try to suppress these terrifying thoughts, she had developed a ritual of walking forward in precisely measured steps and then retracing these exact steps backward. She realized this was not rational, but she thought that walking backward would somehow undo the horrific visions of her parents in danger.

"I told her that we all have terrifying thoughts, and that it doesn't mean you are crazy," Rao said. "And I explained that her precision and analytical abilities had become diverted to these irrational interior thoughts. We worked to redirect this ability outward — back to music and to math."

The girl was weaned off medication and underwent behavioral therapy, focusing on the very thoughts that she found so disturbing. When her anxiety reached its apex, Rao coached her to wait a few moments before retreating to her ritual. This process was repeated, each time increasing the delay between the disturbing thoughts and the walking compulsion. Eventually, the debilitating cycle was broken. Now, when Jan feels anxious, she practices the technique that Rao taught her. Her schoolwork is again outstanding, and she continues to play the piano. Some studies show that such behavioral therapy can be as effective as medication in overcoming obsessions and compulsions.

On the other hand, Rao pointed out, distress is an unavoidable dimension of human experience. "Struggle over suffering and pain is necessary for development," he said. "If you ignore this, or try to medicate it away, then a person doesn't develop skills to deal with life."

In order to better understand a different point of view, I sought out Dr. Joseph Biederman in his office at the Massachusetts General Hospital. He is a professor of psychiatry at Harvard and chief of the joint program in pediatric psychopharmacology at Massachusetts General and McLean Hospital.

Biederman rejects the contention of behavioral psychologists like Anthony Rao that children and adolescents are overdiagnosed because of the exigencies of HMOs or the incentives of drug companies. Nor does he think that overzealous teachers are to blame. "The schools are not failing when they insist that a child cannot endlessly obsess over some task," he said. "That's the response of a classically narcissistic parent — the child is an extension of himself, and it's the environment that must change."

Biederman also finds this response to OCD naive. "These are disruptions of normal brain functions — the diseased limbic loops are organic, not philosophical. It's a pathological state." It was not to be confused with, say, the behavior of a basketball player who mumbles things to himself before a free throw, or a pitcher who wears a special undershirt on the mound. "That's primitive, magical thinking that doesn't really interfere with functioning." Biederman illustrated his point by likening OCD to high blood pressure or high cholesterol: there are well-defined limits beyond which disease occurs.

"As a doctor, you see a patient with high cholesterol, and you tell him to lose weight, exercise, restrict his fat intake. Rarely can anyone do this," he said. So a doctor will prescribe medication. "Do you want to walk from Boston to New York City, or take a plane? Behavioral therapy is the most laborious form of treatment — it's walking. Medication gets you there quickly." He went on, "What we need is a screwdriver for the brain, in order to fine-tune the limbic circuit just enough so that it works efficiently and doesn't get stuck."

But isn't that measurement of efficiency a highly subjective one? "Even minor illness deserves aggressive treatment," Dr. Biederman replied. "Treat early, at the first sign, when a person is still functional. Those early indicators are what I call kindling — you want to intervene when the fire is just beginning, not let it spread." After all, what he and the psychologists had in common, he said, was the goal of alleviating pain. "These people with OCD are suffering, experiencing distress, in a state of hyperarousal. Without drugs, they can't enjoy life."

I also spoke to Dr. Judith Rapoport, the chief of child psychiatry at the National Institute of Mental Health, in Bethesda, Maryland. Dr. Rapoport, who helped bring OCD to the public's attention eleven years ago with her book *The Boy Who Couldn't Stop Washing,* emphasized that exact terms should be used in any discussion of the disorder. "I see the kids whose lives are wrecked by intrusive thoughts and uncontrolled compulsions," she said. "The *DSM* criteria are carefully constructed around degree of function.

"Your scientist friends are not the kind of people referred to me," Rapoport went on. She believes that these scientists more closely resemble an alternative *DSM* diagnosis, obsessive-compulsive personality disorder (OCPD). The *DSM* criteria here emphasize excessive devotion to work and perfectionism. A person with OCPD relentlessly engages in work, to the exclusion of social pleasures.

Other experts in the field believe that the lines are more difficult to draw. Dr. John Ratey and Catherine Johnson, in their 1997 book, *Shadow Syndromes,* argue that many obsessive people do not clearly fit *DSM* criteria for OCD or OCPD. Further, of all the shadow syndromes, "mild obsessive-compulsive disorder is perhaps the one constellation of mood and thought society cannot do without." What links the compulsions is the fear of shame upon failing in public — the scientist who emerges from his laboratory to present his data, the pianist at each recital in the concert hall. "This is the intersection of shadow syndrome and normalcy," Ratey and Johnson argue. "Obsession can drive ambition — and when it does, obsession becomes a useful quality to possess."

Late one evening, I called Laurence Lasky, a nationally renowned molecular biologist, to discuss my findings with him. I was not surprised to find that he was still in his lab at Genentech, a biotechnology company in the Bay Area. Lasky agreed that laboratory researchers, himself included, exhibit traits that are distributed toward the far end of the bell-shaped curve of obsessions and compulsions. "As an adolescent, I had this compulsive habit of tapping and drumming on tables, walls, my books," he told me. "It drove my mother crazy. I needed to do it to burn off my anxiety. As I grew older, it went away." Lasky still lives with anxiety, but he said that there is a huge payoff from all this tension.

"There is nothing like making a discovery — the feeling of seeing something in the laboratory for the first time," Lasky said. "But no good scientist I know is ever completely satisfied. What does a Nobel laureate do when he wins? He tries to win again. If you're not first, and you're not right, you're nowhere." I asked Lasky if he would want to be medicated for his anxiety. Absolutely not, he replied. "Who says advancing science has anything to do with being happy?"

STEPHEN S. HALL

The Recycled Generation

FROM *The New York Times Magazine*

ALMOST EVERY WEEKDAY MORNING, usually before 10:30, an overnight delivery truck with an unusual cargo negotiates the hilly streets on the outskirts of Worcester, Massachusetts, and comes to a halt in front of a brick and tinted-glass building called Biotech Three. The courier disappears into the building with one or two large gray containers and drops them off at a small company called Advanced Cell Technology. The gray cases look like toolboxes, but they are actually sophisticated shipping containers, commonly used for transporting materials used in animal-breeding work, and they contain the starting material for a series of experiments that may completely rewrite the tables of human longevity. Or they may be remembered only for being among the most ethically troubling scientific endeavors of our times.

Inside the gray cases are hundreds of cow eggs — big, plump and beautifully rotund oocytes, as they're technically known — each one painstakingly plucked the day before from the ovaries of cows slaughtered in Iowa. They are doused with a marinade of enzymes that prime them for imminent fertilization and then sealed in small plastic test tubes before being shipped overnight to Worcester. On the drizzly, overcast day in mid-December when I visited the laboratories at Advanced Cell, the shipment of eggs arrived a little before me. In all the time that cows have roamed the planet, their oocytes have never encountered the insults they were about to face that day. The eggs would be stripped of their DNA, deliberately fused with human cells, and fooled into thinking that fertilization had occurred, in the fervent commercial hope that some sort of an embryo might result.

Experimenting with embryos created from two different species — to say nothing of engaging in a form of human cloning — is an enterprise fraught with scientific and social uncertainty; indeed, Congress forbids the National Institutes of Health to finance any research involving human embryos. So it's natural to wonder why a small, largely unknown and understaffed biotech company would risk public scorn and ethical outrage to perform such research. One possible reason is publicity, which a number of critics have been eager to suggest. But the real answer, insists Michael D. West, the dreamy-eyed forty-six-year-old entrepreneur who heads ACT, is the scientific chase for what he calls "the mother of all cells — the embryonic stem cell."

The embryonic stem cell is an almost mythically powerful and versatile human cell, fleetingly present during the earliest days of embryonal development. This one cell has the potential — the genetic blueprint and the biological know-how — to become any cell, any tissue, any organ in the human body. With the proper biochemical coaxing, for example, it can turn into heart muscle, which could replace tissue damaged by a heart attack. Or into brain cells, which could be used to treat Parkinson's disease. Or into retinal cells, which could be used to restore failing vision.

Imagine, in short, a cell so protean and potent that it could theoretically generate an infinite supply of replaceable body parts — organ and skin, sinew and bone, blood and brain — to knit the tatters of disease, injury, or old age. Imagine further that, with the use of controversial technologies like cloning, you might one day donate a snippet of your own skin, allowing scientists to harvest stem cells that theoretically would become a self-generated and limitless supply of transplant tissue — tissue that would make a perfect immunological match with you because, after all, it is you. These ideas have not only been imagined; patents and licensing agreements are already in place.

"These are incredible cells," gushes Thomas B. Okarma of Geron Corporation, the California biotechnology company that controls many of the patents and licenses on stem cells. "I've been in biology since high school, and this is still a chilling technology when you see these things and realize what they do. The number of applications is mind-boggling." Last month, *Science* magazine dubbed stem-cell technology the "breakthrough of the year."

In spite of this promise, the scientific, commercial, and ethical landscapes that intersect with these fascinating cells have been shaped, even turned upside down, by the right-to-life debate, because in order to obtain human embryonic stem cells by currently available technology, you must destroy a human embryo. Because of the congressional ban on financing of this research, many scientists now find themselves in the awkward position of possibly losing their jobs if they try to use human stem cells to devise new treatments for many common diseases. "With stem cells," says Ronald McKay, a researcher at the National Institute of Neurological Disorders and Stroke in Bethesda, "you tickle them and they jump through hoops for you." McKay's group published a highly regarded paper in *Science* last summer showing how rat embryonic stem cells could be used to treat a version of multiple sclerosis in rats, but he doesn't dare extend the research to test the method's effectiveness in humans. "I would get fired if I did that," McKay says, fairly squirming in his seat to make this seem like the most reasonable thing in the world.

The ban on NIH financing has also had the collateral effect of relegating the technology to the private sector, where embryo research can proceed unencumbered. But the fact that so much of this controversial research now occurs in the private sector means that public discussion has become constrained. And complicating this commercial landscape is the fact that two of the companies most avidly competing over the research, Geron and Advanced Cell Technology, have a tangled corporate history dating back to business opportunities created by the federal ban.

The common figure wandering through all these unsettled landscapes, popping up again and again like some white-coated Zelig, is Michael West. West's insistence on pursuing controversial experiments in which cow eggs are used as cellular incubators in an attempt to create humanlike embryos comes at a particularly sensitive moment. Emboldened by a legal interpretation that seemed to show a way around the congressional ban, the NIH on December 2 issued long-awaited guidelines explaining the rules by which the agency would finance human stem-cell research. The sixty-day comment period ends tomorrow, and the agency hopes to begin soliciting grant proposals from university researchers sometime this summer.

However, Representative Jay Dickey, a Republican from Arkansas who has blocked NIH financing for embryo research since 1995, has already vowed to outlaw federally financed stem-cell research this spring, either through legislation or legal action. Senator Arlen Specter, meanwhile, who is entranced by the medical possibilities, has promised to introduce a bill explicitly allowing federal financing for the research.

West, whose laid-back, soft-spoken demeanor belies the soul of a headstrong provocateur, promises to be an impassioned voice in this debate, too. "I think a lot of the problem we have in trying to develop these new technologies for medicine is people's knee-jerk reaction to words like 'fetal' and 'embryo,'" he says. "You know, you use the word 'fetal' and people just go completely irrational."

Until we learn how to direct the fate of human embryonic stem cells — learn how to tell them, for example, that we'd like them to become a new liver or glistening white bone — their enormous promise still remains under Nature's lock and key. Many scientists believe it will take at least ten years before we learn how to program the development of cells that could be transplanted into humans, and probably many more before we learn how to coax stem cells into creating something as grand and complex as a liver or a kidney.

But let's say some version of the science will be in practice by the year 2011, when the first of the baby boomers turn sixty-five. And let's say the technology of stem cells will be complemented by promissory notes redeemed in related fields of gerontological science. What might the menu of spare parts look like?

The first step in creating those spare parts (which could be used not only for the aged but also for anyone in need of intervention) might well be the donation by the patient of a biopsy sample, which could be quickly cloned, thus creating an early embryo, which would produce stem cells in a week or so. Every cell and every tissue derived from those stem cells would be a perfect immunological match, which would immediately circumvent the big stumbling block of current transplant medicine: matching tissue.

At that point, it simply becomes a matter of matching the ailment to the right cell type. Researchers speak of creating nerve cells to treat spinal-cord injuries, stroke, and Alzheimer's disease; glial cells for multiple sclerosis; pancreatic islet cells to treat diabetes; muscle cells for muscular dystrophy; chondrocytes for arthritis; hepatocytes for cholesterol metabolism; endothelial vessels to grow new blood vessels to replace vessels clogged by fat and plaque — there are more than 200 different cell types in the body, and stem cells can theoretically be nudged to form each one. If these cells are souped up, as has been proposed, with an enzyme that maintains its cellular youthfulness, we're talking not only about replacement parts, but also about parts that never grow old.

What might the world of stem-cell medicine look like? For one thing, every place in the developed world might look a lot more like Florida. Although the maximum attainable human life span is now approximately 120 years, only about 65,000 Americans have currently reached the age of 100. But a century from now, with new medical technologies in place, the Census Bureau predicts there will be 5.3 million people living to the age of 100 and perhaps much longer. To hear the optimists talk, there'll be much longer waits for tennis courts and tee times.

This is the kind of speculation that keeps ethicists and philosophers busy, but there are a lot of touchier issues on the immediate horizon — creating embryos for research purposes, combining biological material from different species, and performing nuclear-transfer experiments involving adult human cells (also known as cloning). All three are being pursued in a single line of research at Advanced Cell Technology.

The experiments at ACT take place in a long, dark, narrow room in the middle of the lab. The lights are kept low to allow technicians to view the cells through microscopes while poking, turning, and prodding them with micromanipulation devices. K. C. Cunniff, her blond hair spilling out over a white lab coat, begins by vacuuming all of the cow's genetic material from the egg cells, one after another, for more than an hour.

As I watch a magnified view of the proceedings on a television monitor, it's hard not to be impressed by the nimble piecework of the technicians. Cunniff, twenty-six, deftly maneuvers one pipette to hold a round, plump egg cell in place. (They prefer cow eggs because they are inexpensive and easier to use than either pig or primate oocytes.) She turns the egg, which had previously been stained with a fluorescing dye, as if rotating a ball in her hand, until she sees a telltale blue-green speck of DNA. Using her other hand, she nudges a sharp, hollow needle up against the surface of the egg. With a precise thrust, she enters the egg, sucks out the DNA, and withdraws the needle, all in a matter of seconds. "Very quick," she says matter-of-factly. In, suction, out. Next egg. In, suction, out.

It's so matter-of-fact that you'd never guess this type of experiment was the focus of wildly fearful predictions only a few years ago. In 1997, when scientists at the Roslin Institute in Scotland announced the cloning of Dolly the sheep, there was a lot of fevered speculation about the possibility of cloning human beings. It was generally dismissed as both a distant and undesirable proposition, but the sudden appeal of embryonic stem cells has provided a surprisingly urgent medical justification for moving the technology of human cloning closer to reality.

In the case of ACT's experiments, the cow egg is being tested as a way to "reprogram," or reset, the adult DNA back to a pristine state resembling the moment of fertilization. In these experiments, the egg comes from a cow but the adult DNA comes from human donors; it's the human DNA, West believes, that orchestrates subsequent embryonic development. The technical term for this type of experiment is "nuclear transfer," but West prefers a more transparent phrase — "human therapeutic cloning." That modifier "therapeutic" signals that ACT is not in the business of making babies, just embryos to get stem cells. And they're not alone in pursuing the nuclear-transfer approach. Last May, to remarkably little public comment, Geron acquired Roslin Bio-Med, the Scottish company in charge of commercializing the research that produced Dolly.

As Cunniff moves another cow egg into position, Nancy Sawyer gazes at the TV monitor and makes a sound like "Unk!" as the needle punctures the egg. "That's so cool," she whispers of the image. Seen on the monitor, the image has a primal iconic beauty, like a dim, grainy photograph of a distant planet. On each moonlike oocyte, one isolated lunar mare glows with an eerie greenish blue, betraying the location of genetic material that will soon be suctioned away.

At a nearby microscope, Sawyer, twenty-four, performs the next step. Human skin cells, known as fibroblasts, have been provided by an anonymous donor. The human cells dot the screen like so many lumpy, transparent potatoes, much smaller than the egg cells. Sawyer loads them into a hollow needle, immobilizes a cow egg with suction, and, with gentle pressure, inserts the needle tip just under the rind of the egg. With a squeeze of her hand, a single human skin cell, with its unique payload of human DNA, plops into the egg of a creature that moos. Sawyer even manages to maneuver her needle tip to literally tuck the hole closed.

After stuffing every cow egg with its little spud of human DNA, Sawyer prepares the next step. She gives the cells a zap of 120 volts. The jolt of electricity effectively fuses man and beast into a single biological fate. After one final step, this . . . this thing will believe it has been fertilized and, if all goes well, begin cleaving, or dividing, in the bubbling, momentous arithmetic of life lifting off the pad: 2 cells, 4 cells, 8 cells, 16 cells, 32 cells — blastocyst!

The odds are extremely long that any of the 100 or so cell fusions will result in a blastocyst, the hollow ball of cells that appears about a week after fertilization and, in a normal embryo, is destined to become the placenta. But if that ball of cells should form, a separate group of cells known as the inner cell mass assembles against one inside wall of the blastocyst, like bats huddling on the ceiling of a cave, and this is where the embryonic stem cells will appear. These stem cells exist only briefly before they begin to differentiate into more specialized tissues — there one moment and gone the next, transients vanishing into the great biology of becoming. "If I'm lucky," says Jose Cibelli, vice president of research at ACT, "I'll get one blastocyst. That's how low the efficiency is." And getting the blastocyst is just the first hurdle — can they isolate the stem cells and keep them alive in a test tube? And will they truly act like human embryonic stem cells?

Before leaving Worcester that day, I ask Cibelli when he will know if any blastocysts have formed from the experiment I've observed. He says he wouldn't expect to see anything before nine or ten days. I do the math in my head and make a mental note to get in touch on either December 24 or Christmas Day.

In many ways Michael West is the shadow impresario of the field. As founder of Geron Corporation, one of this decade's most closely watched biotechnology companies, and now as president and CEO of Advanced Cell Technology, West has achieved remarkable success as a kind of merchant of immortality, selling the idea that stem cells and related technologies might someday completely revise the tables of average human life span. And he is so convinced that the promise of stem cells justifies a controversial strategy like cow-human nuclear transfer that he is happy to foster, if not force, a national discussion of this technology.

In a recent issue of the journal *Nature Medicine,* for example, West, Cibelli, and their colleague Robert Lanza argued the case for human therapeutic cloning because stem-cell research is so promising. "Does a blastocyst," they wrote, "warrant the same rights and reverence as that accorded a living soul — a parent, a child or a partner — who might die because we failed to move the moral line?" And that is what Mike West is trying to do at this touchy juncture of the stem-cell wars — with this company, in these experiments, in an ethical debate that has seemed too arcane and complicated to attract much public attention to date. He is putting his shoulder to the moral line that forbids embryo research and is trying to force some sort of social reckoning. The degree to which he and others succeed or fail may well determine if stem cells have a chance to live up to their promise as medicine or remain too hot to handle in the current political climate.

Stem cells burst into public consciousness little more than a year ago. In November 1998, James Thomson of the University of Wisconsin reported the creation of human embryonic stem (or ES) cell lines. Using leftover frozen embryos from in-vitro fertilization clinics in Wisconsin and Israel, Thomson and colleagues isolated stem cells and have shown that they can be maintained indefinitely in the lab — can be grown, frozen, and then thawed, and still retain their power to develop into, say, heart-muscle cells or brain cells. Michael Shamblott and John Gearhart, at the Johns Hopkins School of Medicine, headed a separate effort to cultivate something called "embryonic germline" (EG) cells, which are harvested from a tiny speck of fetal tissue from an aborted fetus and then grown in the lab. Because of the congressional ban on federal financing for human-embryo experimentation, both teams conducted the research with financing from Geron.

There are many technical hurdles to overcome. But the sheer power of the approach makes it clear that, if properly harnessed, stem cells could serve as a warehouse of spare human parts. This teasing hint of immortality is the cultural subtext that runs beneath the public fascination with the science, and no one has done more to promote that connection than West. The son of a truck mechanic, West is a one-time creationist and a self-styled truth-seeker, and his entrepreneurial interest in the biology of aging derives from an obsessive, almost morbid fascination with death. "All I think about, all day long, every day, is human mortality and our own aging," he says.

The first time I visited Advanced Cell Technology, West showed up two hours late for our appointment, apologized with sheepish charm for his tardiness, and began to spin out the kind of polished futurism he regularly conveys to scientists, investors, and laypeople. "I thought I'd show you some pictures," he said, and then proceeded to project slides on a screen in the company's conference room, delivering a lecture on the biology of aging to an audience of one. With his gently soothing Midwestern voice and relentlessly upbeat brand of biological positivism, he manages to make science sound almost like a cult. Former colleagues concede that he does not possess a crisp management style (punctuality, for example, being a continuing challenge), but even his critics admit he has a knack for looking beyond the horizon and dreaming deep. He'll talk for hours about saving endangered species through cloning, or the possibility of cloning pets, or why it makes more economic and ethical sense to pay $1 for a cow egg than $2,000 to surgically obtain a single human egg. "He's a very visionary guy," says James Thomson, a University of Wisconsin biologist, "but he's also a very good salesman."

Even as a high school student growing up in Niles, Michigan, West was fascinated with aging and rejuvenation. He immersed himself in religion and philosophy. He learned Hebrew and Greek, he says, to read ancient texts in the original. He went on to study psychology at Rensselaer Polytechnic Institute, obtained a master of science degree at Andrews University, a Seventh-day Adventist school, and even studied creationism for a time in San Diego, he says, before convincing himself of the truth of Darwinian evolution. Following the death of his father in 1980, West worked in the family's truck-leasing business and then belatedly embarked on a career in science. He received his Ph.D. in cell biology from the Baylor College of Medicine in 1989 and started medical school at the University of Texas Southwestern Medical Center in Dallas.

That same year, West showed up unannounced and began to hang out at the University of Texas lab of Woodring Wright and

Jerry Shay, who were researching the molecular biology of aging. Four years earlier, scientists had discovered a critically important enzyme called telomerase, which acts on telomeres, the little caps of DNA at the end of chromosomes; telomeres ordinarily grow shorter each time a cell divides, until the cells stop dividing altogether. It turned out that a handful of human cells — germ cells, cancer cells, and embryonic stem cells — use telomerase to circumvent that shortening process, and thus also circumvent the aging process and achieve a cellular version of immortality.

As West learned more about telomeres, he eventually came to view the enzyme telomerase as a molecular version of the fountain of youth; it looked as if it might bestow immortality on normal cells and, conversely, could have the beneficial effect of pulling the plug on immortality in cancer cells if blocked. While still technically enrolled in medical school, West moved out to California and banged on doors in search of seed money for a biotech startup. Thus, in November 1990, he founded a company dedicated to the molecular causes of aging. He named it Geron — Greek for "old person." He eventually captured the interest of the most prestigious venture capital firm on the West Coast, Kleiner Perkins Caufield & Byers, which along with other firms invested $7.6 million in Geron in 1992.

For a company that has lost tens of millions of dollars and is in no danger of curing aging anytime soon, Geron has managed to cast a spell on investors, the media, and the lay public. In both technical articles and news releases, it has retailed a scientific vision (and vocabulary) that clearly pushes the right Zeitgeist buttons. West and Geron spoke tirelessly of "immortalizing enzymes" and the "life extension" of cells; Geron is almost universally recognized as an "anti-aging" company. And in 1997, after winning a highly competitive race to clone (and patent) the human gene for telomerase, Geron actually had a real molecule around with which to develop clinical products. The company currently has a number of promising directions for telomerase-based products, including potential anticancer applications, but the mythology is so firmly established that even though company officials insist Geron is no longer an "anti-aging" company, this is still how it is inevitably portrayed.

From the very beginning, however, West had another big idea he wanted to pursue. "You need replaceable cells and tissues for the problems of aging as well," he said. "And it seemed to me that the ideal source for an aging population is to go back to the beginning of life." To the embryo, that is, and stem cells. And so, as early as 1992, he paid a visit to Roger Pedersen, a professor of obstetrics, gynecology, and reproductive sciences at the University of California at San Francisco, and a leading expert on embryonic stem cells in mice. They both agreed that the time had arrived to explore the vast potential of such embryonic stem cells in human medicine, and West inquired into whether Pedersen would accept financing from Geron to do research on human stem cells.

Pedersen flatly refused: "This area of investigation is something that is at the headwaters, and it's not appropriate for private investors to control the headwaters of a stream of research."

Several years later, however, Pedersen got back in touch with West. Circumstances had changed, he said, and he was ready to deal.

The circumstances were political. In 1975, federal regulations stipulated that any government support for in-vitro fertilization research required the approval of a federal ethics advisory board. After a short and turbulent history, this board was disbanded in 1980 without a single research project having received government funds. One cynical stratagem of the Reagan and Bush administrations, according to Pedersen, was to block all attempts to reconstitute the panel, effectively thwarting such research throughout the 1980s.

Under the new Clinton administration, Congress did away with the phantom federal ethics review and set up a special NIH committee to establish guidelines for human embryo research. Everything seemed back on track when, early in 1995, Jay Dickey, the Republican congressman from Arkansas, successfully inserted a rider into the budget bill for the Department of Health and Human Services (which includes the NIH), banning federal funds for human embryo research. Just as right-to-life politics had forced in-vitro fertilization and reproductive biology into the private sector, where lack of oversight and regulation has led to a series of well-documented scandals, stem-cell research seemed headed for similar privatization.

Those were the circumstances that had changed Roger Pedersen's mind. Geron reached an agreement with the University of

California to finance Pedersen's work on embryonic stem cells. During his discussions with West, Pedersen mentioned that a scientist at the University of Wisconsin, James Thomson, was about to publish a paper announcing another breakthrough: the isolation of embryonic stem cells from rhesus monkeys. West was in Wisconsin the next day, and Geron signed up Thomson too.

"I would have been much happier with public support," Thomson admits. "But given the constraints, I welcomed the funding I got." Getting access to stem cells from a primate, an animal biologically close to humans, West realized, promised tremendous intellectual-property dividends. "We could just learn how to work with them, and file patents," he said. "But we'd have this head start on the whole world." When West later learned that another university researcher, John Gearhart at Johns Hopkins, had made significant progress isolating cells very similar to embryonic stem cells, he headed straight for Baltimore. "He just showed up on my doorstep one day," Gearhart recalled with a laugh.

With Pedersen acting as talent scout, West chased down and signed up three of the leading stem-cell researchers in the world. Geron began to assemble a staggering intellectual-property portfolio in a field with almost limitless medical potential. And because the congressional right-to-life advocates had effectively tied the NIH's hands in terms of financing, there wasn't any competing research in university labs. The investment paid off spectacularly in November 1998, when both Thomson and Gearhart announced they had isolated human stem cells. The universities where the work was done retained the patents on the research, but Geron received exclusive rights to many applications.

Unfortunately, Mike West enjoyed this moment of triumph only vicariously, because by then he had left the company, unhappily. In 1997, according to several sources, Geron planned to spin off its entire stem-cell program into a separate company, which would include West, when the plan, in the words of a scientific board member of the company, "got clobbered by the company leadership." Thomas Okarma, who joined Geron in December 1997 and is now president, offers a different interpretation. "I was hired explicitly to run that program because it really wasn't moving," he said. In any event, the stem-cell research stayed at Geron, and West says he increasingly felt he could do more outside the confines of the company than inside.

The fact that West's current company, Advanced Cell Technology, is now competing against the company he founded may go a long way toward explaining why Mike West is so determined to find an alternative way of obtaining human stem cells, one that doesn't rely on existing human embryos in clinics or fetal material from abortions — the methods that Wisconsin and Johns Hopkins licensed to Geron. And it certainly explains why he perked up when, a few months after leaving Geron, he learned of an unusual experiment by Jose Cibelli, an Argentinian scientist working at Advanced Cell Technology. The good news was that Cibelli had tried to isolate human stem cells using a method that seemed to offer an alternative to Geron's approach. The bad news was that the strategy involved human cloning. And cows. "And I knew," West says, "that was going to be a problem."

In the summer of 1996, Cibelli was a graduate student at the University of Massachusetts at Amherst, working in the laboratory of James Robl, a respected developmental biologist. Cibelli had the radical idea of fusing some of his own cells with cow egg cells, in effect cloning himself — not to make a copy, of course, but as a way to get human stem cells. Interestingly, this wasn't the first time Robl had been confronted with such an idea. Several years earlier, a student in the lab, unbeknown to Robl, had fused human cells with the egg cells of a rabbit, and the cells had begun to divide. This time, Robl went to university officials and received institutional approval to proceed.

During July and August, Cibelli rinsed out some of the cells that lined the inside of his cheeks and tried fusing them with cow egg cells. "One day, Jose was about to go on vacation, and he was about to throw out the dish," Robl recalls. "I don't generally look at these things, but I did that day. And there was a blastocyst." In other words, the embryolike thing had moved beyond mere cell division and graduated to the stage where embryonic stem cells begin to form. As is routine when the experiment reaches this stage, Cibelli placed the fragile blastocyst on a bed of fetal mouse cells, which nourished its further development. "We watched it for about two more weeks," Robl says. "It looked, to my eye, not like a cow blastocyst. The morphology of the cells was different."

Cibelli and Robl did not publicly discuss the experiment at the time, nor did they prepare a scientific report for peer review. "We never considered a publication," Robl explains, "because there was not nearly sufficient data." But the University of Massachusetts did consider the experiment sufficiently novel to file a patent application. The United States patent was issued, virtually unnoticed, last August.

Cibelli recounted this remarkable story to West in the spring of 1998, when West happened to be visiting ACT. "I was just flabbergasted," West recalls. "I mean, he showed me human embryos that had been made by cloning. And I had no idea — no one in the world had any idea — that it had been done. I thought, 'Oh my gosh, this is exactly what I want to be doing for the next ten or twenty years of my life.'" More to the point, it was exactly the kind of technology that would allow him to get back into the stem-cell game.

West officially joined ACT in October 1998 and immediately presided over an episode that was, for him, uncharacteristic — a major public-relations fiasco. As soon as West joined the company, Cibelli lobbied to resume his cloning experiments. West agreed, but wanted to disclose details of the 1996 experiment and gauge public reaction to the technology before starting it up again. "I didn't want to be accused of doing this in secret," he says. So he invited a film crew from the CBS newsmagazine *48 Hours* to film the work in progress.

Then, one week before the scheduled broadcast, on November 6, West got blindsided by his previous life. Geron announced Thomson and Gearhart's successes isolating stem cells. "I had no idea it was coming," West admitted. It wasn't just that the research made front-page headlines and drove Geron stock up 74 percent in one day. By the time CBS broadcast the show on November 12, along with a news account of the experiments that appeared in that morning's *New York Times,* West's professed desire for openness looked like something entirely different: a bid to leverage "me too" publicity for his otherwise unknown company.

For an experiment that never received formal peer review, Cibelli's cow-human nuclear-transfer work got plenty of unofficial feedback, beginning with the White House. Clinton called it "deeply troubling." Thomas Murray, president of the Hastings Center and a leading bioethicist, wondered "if the timing of the announcement had to do with scientific competition, personal competition or positioning for funding from investors." Roger Pedersen was quoted as saying, "I smell a sham." (He claims he wasn't told who performed the experiment when he was asked to comment upon it.) Thomson regarded the whole affair as "unfortunate." Right-to-life pickets showed up outside Biotech Three in Worcester, marching around in cow masks.

Two days after news of the cow-human experiments broke, President Clinton asked his National Bioethics Advisory Commission to prepare a report on stem-cell research in general. The commission hastily convened hearings in January 1999 in Washington. And who should turn up, uninvited, to make an unscheduled presentation before the panel but Michael West.

"I read an editorial by an individual who wrote that science should stop so that ethics can catch up," West told the national commission that day. His ambition, he said, was "to communicate to people in public policy and in biomedical ethics, so that, simply, ethics can walk hand in hand with science." Ethics and modern biology, however, have rarely walked hand in hand, and stem-cell research has added new difficulties to the relationship.

Over the past year, two high-powered advisory committees have, with one significant difference, endorsed the general idea of embryonic stem-cell research. A committee established by the American Association for the Advancement of Science recommended last August that researchers be able to receive public funds for experiments on embryonic stem cells, but only using cell lines already created by researchers in the private sector. The National Bioethics Advisory Commission, by contrast, recommended last September that researchers financed by the NIH should be allowed to create stem cells using human embryos already in existence (thousands of such embryos, frozen and destined to be discarded, exist at in-vitro fertilization clinics).

But no oversight or guidelines exist for private industry, and in February 1999, well before either committee delivered its opinion, Advanced Cell Technology quietly decided to resume its cow-human nuclear-transfer experiments. It seemed like an important decision for such socially sensitive research, so I asked West if it was something that went to the board or an ethics advisory panel for approval. West said it was strictly his decision. And there's the paradox. Now that abortion politics has forced so much of the research into the private sector, the transparency of the ethical conversation about it has become more obscured.

Everyone, including Mike West, insists on an open national debate about stem-cell research. But I began to notice that whenever I asked one question too many about exactly what work was being done, or even contemplated, the conversations became elliptical and vague. I was having lunch with West one day, for example, when I asked if the use of human donor eggs for cloning experiments was under consideration. Cibelli had told me he thought such a development was very likely. "I have to confer on this issue," West replied apologetically, leaning over to huddle with ACT's public-relations adviser. Then, after a pause but without directly answering the question, he expressed concern that such a program could exploit women.

For all their good intentions, ethicists may have allowed themselves to be placed in a difficult, possibly untenable position. At Geron, for instance, the company's ethics advisory board seems to have the ear of management. But, says Karen Lebacqz, who heads the Geron ethics advisory board, "they are perfectly at liberty to ignore all our advice." Further, as Lebacqz points out, she is not free to discuss certain aspects of the research. "Early last summer, they brought us a piece of research that they were going to fund. Several members of the board raised objections, so they decided not to pursue that particular line of research." What was the research under discussion? "I'm sorry, I really can't," she said.

The ethicists have become proxies for all of us, precisely because so much of this technology, for political reasons, is unfolding in the private sector. Yet they have limited power. West, for example, told me that when Geron first considered establishing an ethics board, the company determined that giving such a board the right to veto research projects would undermine its fiduciary responsibilities to shareholders. The ethicists have what West calls "the power of the pen," but what they can report back to us is constrained.

Their mere participation in the process, however, creates the appearance of oversight and ethical responsibility, and that is precisely what bothers David Cox, vice chairman of the genetics department at Stanford University and a member of the national bioethics commission. Cox says ethics advisory boards at biotech companies are "a joke": "They're supposedly doing ethical review, but the process by which they're working is backwards."

It's very hard to have a national debate on issues as socially and ethically important as cloning and the creation of embryonic stem cells when every conversation may ultimately bump up against corporate confidentiality. The problem of openness is compounded by the editorial policy at a number of leading scientific journals, which refuse to publish research if the results have previously been disclosed in public. That almost guarantees that breakthroughs in a controversial and competitive field of research like stem cells will land in the public's lap as scientific faits accomplis, just as Dolly did.

And it leaves us in the same scientifically uncertain and ethically queasy place we were more than a year ago. At that time, Thomas Murray of the national bioethics commission asked West if he thought the cow-human experiments resulted in human embryos that were potentially "viable" — in other words, embryos that, if implanted in a woman, could result in a live human being. West didn't answer, and ACT still isn't answering. I asked Jose Cibelli, for example, if the ACT scientists had made any progress overcoming the problem of biological incompatibility between a cow egg and a human cell, an issue involving small cellular organs called mitochondria; it is an obstacle repeatedly raised by the many scientists who remain skeptical about the approach and are dubious that a blastocyst would be created, especially since ACT still hasn't published a sprig of data on it in more than a year.

"We think," Cibelli began to say, "but this is very preliminary . . ." He shrugged and smiled. "You can say that we've had good progress," he continued with a little laugh. "We've had good progress, and we expect to have something to report in the near future. But I guess I need to be protective of the data for publication's sake."

Michael West sounded a similar theme when I told him how many complaints I'd heard about ACT's unpublished experiments. "We could publish now, the data we have," West assured me, well aware of the exasperation of the research community. "But," he continued, "we're trying to generate a real killer paper here. We're going to do a paper we're proud of."

On Christmas Eve, I sent an e-mail to Jose Cibelli, asking about the status of the cells I'd seen fused. The day after Christmas he wrote back to report that the nuclear-transfer units had "developed at the 'predictable' rate," while declining to specify exactly what that was. As soon as new techniques were in place, he continued, Advanced Cell Technology would report the results in a peer-reviewed journal.

As Congress prepares to debate the merits of stem-cell research in the coming months, we will undoubtedly hear rosy visions of the future of medicine with stem cells (as well as the contorted political logic that suggests that research on human embryos and cross-species nuclear-transfer experiments are permissible in the private sector, but morally indefensible in the public sector). But it's also worth thinking through the implications of immortalizing medicines, which I had the opportunity to do with Leonard Hayflick, the elder statesman (and elder contrarian) of the field.

Hayflick, now seventy-one, is a well-known cell biologist; his discovery in 1961 that normal human cells grown in a test tube simply stop dividing after a specific number of cell divisions, known as the Hayflick limit, in effect introduced the notion of mortality into the biology of aging. We arranged to meet at the Union Club in Manhattan in early December, when Hayflick showed up to harangue fellow board members of the American Federation for Aging Research about their financing priorities.

Hayflick doubts we'll ever have a quick fix to arrest aging anytime soon, but he has lots of reasons to think it would be a very bad idea. Like a number of bioethicists, he believes the first line of division is economic. Access to the regenerative medicine of stem-cell and immortalizing enzymes is most likely to be a phenomenon available only to affluent segments of the population in the developed world.

In his book *How and Why We Age* and other writings, Hayflick has even gone to the trouble of imagining "bizarre situations" that might unfold if scientists were ultimately able to create a medication that would, from the moment treatment began, essentially freeze the process of aging at a certain point. He has imagined "children becoming biologically older than their parents" if a parent chose to stop aging at age forty-five, for example, while a child did nothing. How would you even know, he asks, the right age to stop at?

Karen Lebacqz has also pondered this distant future. "If we are successful with the use of stem cells or in the reprogramming of cells," she says, "it will mean that people are no longer dying of the things we are dying of today. What do we do with all of ourselves if we don't die?" We'll squander even more of the world's resources, she continues, and put even more pressure on the developing world.

But perhaps the ultimate argument against the implicit promise of immortality has to do with a simple biological fact: if we were to rejuvenate our brains, Hayflick argues, we might lose the most precious thing we have: our sense of self. "Given the possibility that we could replace all our parts, including our brain, then you lose your self-identity, your self-recognition. You lose who you are! You are who you are because of your memory."

There is a lot of scientific research and a lot of heated political debate to come before we arrive at any of those distant quandaries. And how we resolve the ethical conversations we're in the midst of having over stem cells will have a lot to say about whether we'll have a chance of reaching that future at all.

BERND HEINRICH

Endurance Predator

FROM *Outside*

I'M STANDING in an ancient landscape in East Africa. All around me white and yellow flowering acacia trees are abuzz with bees, wasps, and colorful cetoniid beetles. Baboons and impalas roam in the miombe bush. Herds of wildebeests and zebras thunder by; in the distance, elephants and rhinoceroses lumber over the rolling hills like prehistoric giants. Little seems to have changed in the last few million years. Caught up in searching for insects, I happen to peek under an inauspicious rock overhang and am taken aback by what I see.

Painted on the wall is a succession of sticklike human figures, clearly in full running stride. All are clutching delicate bows, quivers, and arrows, and all are running in one direction, left to right across the rock canvas. It's a two- or three-thousand-year-old pictograph, with nothing particularly extraordinary about it — until I notice something that sends my mind reeling: The figure leading the procession has his hands thrust upward in what seems to me to be the universal sign of athletic victory. As both a former ultramarathoner and a biologist, I know this gesture to be reflexive in runners and other competitors who have fought hard and then feel the exhilaration of triumph over adversity.

This happened several years ago, in Zimbabwe's Matobo National Park (formerly Matopos Park), but it remains for me an iconic reminder that the roots of our competitiveness go back very far and very deep. Between the marketing hype, the melodramatic background stories, and the sprawling spectacle of the millennial Olympic Games this September in Sydney will lurk the real reason we will tune in: an intense, innate, even visceral appreciation for the magnificence of the serious athlete's body. The modern Olympics represent the ultimate test of our ancient faculties. We thrill to see athletic skill — abilities that most of us possess to a degree — raised to the utmost level. The Olympics are a product both of our dreams and of our indomitable drive for perfection, the best of what the mortal human body can achieve.

Looking at that African rock painting made me feel that I was witness to a kindred spirit, a man who had long ago vanished yet whom I understood as if we'd talked just a moment earlier. I was not only in the same environment and of the same mind as my unknown Bushman, I was also in the place that most likely produced our common ancestors. The artist had been here hundreds of generations before me, but that was only the blink of an eye compared to the eons that have elapsed since a bipedal intermediate between our apelike and recognizably human ancestors left the safety of the forest for the savanna some 4 million years ago.

It wasn't an easy transition. Indeed, it had fateful physiological and psychological consequences that are still deeply embedded in our bodies and our psyches. Standing before that long-lost victor in the struggle to survive, I was reminded of what I was, still am, and perhaps what we will forever be as long as we are human.

We were all runners once. Although some of us forget that primal fact, comparative biology teaches us that life on the plains generates arms races between predators and prey — and our ancestors definitely weren't into unilateral disarmament. Meat was abundant, for those who could catch it or wrest it from the competition, that is, leopards and lions, not to mention hyenas, jackals, and vultures. Because we primates weren't superb runners, we needed alternatives to sheer speed to eat in the wide-open spaces. So we traveled in groups, racing overland to fresh-killed carcasses and chasing off scavengers. These skirmishes, as well as infighting with our own species — that is, our first true competitors — became the bridge to hunting live prey. The faster you could run, the more valuable you became in the new social groups based on the hunt.

In 1961 I spent a year collecting birds in Africa for Yale's Peabody Museum, and I experienced, I think, what ancient hunters were up against. I'll never forget my feelings of dreary claustrophobia during the months we spent in dense, dripping forests, nor, alternatively, the feeling of glorious exhilaration out on the open steppes. To catch even small birds, I had to wander extensively, half of each day, just as our ancestors must have done. By about 2 to 3 million years ago, they had a leg and foot structure almost identical to our own, and it's reasonable to assume that they walked and ran like we do. While other predators rested, I was able to continue, albeit slowly, because we humans have one major physical advantage: We can sweat, copiously, which allows us to manage our internal temperature and extend our endurance. Most animals have no such mechanism. Through the ages and across the continents there are examples of men actually chasing down beasts that are much faster. In fact, there are modern reports of the Paiutes and Navajos of North America hunting pronghorn antelope on foot, patiently running down a stray till it drops in its tracks from exhaustion and then reverently suffocating the animal by hand.

A quick pounce and kill requires no dream. Dreams are the beacons that carry us far ahead into the hunt, into the future, and into the marathon. We have the unique ability to keep in mind what is not before the eye. Visualizing far ahead, we see our quarry, even as it recedes over the hills and into the mists. Those ancient hunters who had the longest vision — the most imagination — were the ones who persisted the longest on the trail and therefore were the ones who left more descendants. The same goes these days: Human beings with the longest vision tend to make the biggest mark. Vision allows us to reach into the future, whether it's to kill a mammoth or an antelope, to write a book, or to achieve the record time in a race.

Now we chase each other rather than woolly mammoths. But the basic body movements required for hunting and for warfare — running, throwing, jumping — have become ritualized in the track and field events, which are still the heart and soul, the very essence, of the Olympics. The Games are simply mock wars waged in the spirit of camaraderie, though they retain the intensity of their origins. The difference is that in a contest with prey there is always an endpoint: We get it, or it gets away. In our races against one another, in our constant striving to better our achievements and set new records, there is no apparent end. Where, then, are the limits?

*

World and Olympic records have been kept for more than a century, but over that span there never has been a year in which records have not been broken. Performances that were world-class only fifty years ago are almost routine now. Again and again, feats thought physiologically impossible have been surpassed. In 1954, Roger Bannister ran the mile in 3:59.40 to break the four-minute barrier and stun the world. But within six weeks even that improbable mark fell. Fast-forward to 1999 and Moroccan Hicham El Gerrouj lowered the record to 3:43.13.

So it goes: In the Mexico City Olympics of 1968, Bob Beamon shattered Ralph Boston's world long-jump record of twenty-seven feet, four and one-quarter inches with a jump of twenty-nine feet, two and a half inches. For nearly twenty-three years Beamon's record was considered to be beyond unbreakable, until the 1991 World Championships in Tokyo, where Carl Lewis came within one inch of it and Mike Powell actually beat it by two inches at the same meet.

The first modern record for the 100-meter dash was 11.0 seconds, set by Great Britain's William MacLaren in 1867. It got chipped away over the next several decades until American Charles Paddock dropped it to 10.2 seconds in 1921. His time didn't see a major improvement until 1956, when countryman Willie Williams ran a 10.1. Then, last year, U.S. sprinter Maurice Greene set a world record of 9.79.

The steady improvement in records of all sporting events may, at first glance, look like biological evolution, but this couldn't be further from the truth. Evolution might still have played a role shaping us back in the Ice Ages, when we were fragmented into small isolated populations, regularly dropping dead due to athletic deficiencies and other forms of bad luck. No more. Living as we now do, in large, increasingly homogenized populations, any mutation that might crop up and that could be of value for athletic performance (e.g., an enormously large lung capacity for marathoners) would quickly be diffused in the gene pool.

That's not to say changes can't happen. Could a species stuck with our bipedal design evolve and someday run as fast as ostriches? Maybe we're still so unspecialized for the task of running that selective breeding could accomplish this. But even if we attempted that unthinkable experiment — if we bred humans like, say, racehorses, along lines of pedigree — the project would probably have to continue uninterrupted for hundreds or thousands of years. We have no idea what makes a Secretariat different from an also-ran, but if we want to beat a Secretariat, we begin with Secretariat genes. Still, if we did create human thoroughbreds, there's good reason to believe the physical "improvement" would eventually stop; despite selective breeding, thoroughbreds haven't gotten any faster in the last 100 years. Why should it be any different with us?

Genetically we're pretty much the same as we've been for hundreds of thousands of years; the basic changes for running, throwing, jumping, and the like were made long ago, and the trajectory, and eventual endpoint, were determined then as well. Physiologically speaking, on average we may well be *devolving*, so to speak. If we picked one of our 6 billion brethren at random and had that person run against a fit-for-survival Pleistocene man or woman, there's a good chance we'd come out the loser.

Don't tell that to Michael Johnson. To understand performances like his, it's important to recognize that, in terms of genetics, training, and nutrition, a world-record performance is the far, far end of the normal distribution. Olympians don't represent typical physiology. Far from it. World-class athletes are generally off the scale according to every parameter one can think of — physiological systems for muscles, enzymes, hormones, bone structure, and body build. Moreover, all of these superlatives have been bolstered by the best knowledge and execution of diet, rest, training, and stress management. In an Olympic athlete, more and more we're looking at a freak, an elite specimen who is not like you or me and who is fit to do one thing well — likely at the expense of other things.

Each event has circumscribed specifications. For instance, the very best sprinters don't need much aerobic capacity because they rely on a preponderance of fast-twitch muscle fibers, which contract quickly and anaerobically, meaning they don't require oxygen to burn fuel. Those same athletes could not successfully run distance, because long-distance runners rely on a huge aerobic capacity and a larger percentage of slow-twitch fibers, which contract at a slower rate but can work for long periods, so long as they're being continually supplied with oxygen. These traits are largely inherited: If your muscles are made up mostly of slow-twitch fibers, you'll simply never be explosive. We might be able to do a lot to change the basic design we're born with, but not to the point of achieving a world-beating performance.

In the early days of Olympic and world competition, the athletes were probably closer in ability to the average population. Nevertheless, they came from a very small pool out of the total population, and that pool came largely from the privileged class or those who, for one odd reason or another, decided to throw the javelin, long jump, sprint, or run the marathon. Such is not the case now. First, talent is actively solicited: Individuals are identified, nurtured, and encouraged to pursue their dreams to the near-exclusion of more distracting concerns, like milking the cows or otherwise making a living. A second and perhaps much more significant phenomenon is that the pool from which the talented are selected has expanded dramatically. Since 1896, when the first modern Olympics were held, the world population has quadrupled. What's more, while Olympians were previously drawn only from Europe, Australia, and North America, now they also come from Asia, Africa, and South America. Statistically, by simply increasing the sample size, you increase the likelihood of having some individual runner who is faster than ever before in history (as well as one who is slower than ever).

The only real evolution has been in realms not directly related to biology. The most obvious factor in athletic improvement has been better technology. Running shoes are infinitely better. Vaulting poles morphed from ash to bamboo to aluminum to fiberglass, nearly doubling the record heights in the event. And of course, swimsuits have undergone all manner of makeovers, from wool trunks and tops in the early 1900s to skimpy Lycra numbers in the disco years to full-body suits debuting in Sydney called fastskins, which have a dimpled surface, much like a golf ball's, to reduce drag.

Mirroring technological breakthroughs have been changes in technique, such as Dick Fosbury's now-standard backward flop over the high-jump bar and swimmer David Berkoff's dolphin kick in the backstroke. Training methods have also evolved. Germany's Waldemar Gersheler used interval training to help his protégé, Rudolf Harbig, nab the world record of 1:46.60 in the 800-meter run in 1939. Arthur Lydiard of New Zealand helped Peter Snell take Olympic gold in the same event in 1960 and 1964 by advocating long, slow running to build endurance, and brutal hill work to build strength. And Britain's Sebastian Coe, who in 1981 set an 800-meter world record that held for sixteen years, used weight-lifting in addition to Gersheler's and Lydiard's methods.

Such a multitude of factors makes it nigh impossible to predict limits, but physical limits must exist. In just one century the law of diminishing returns has already set in; in certain track events, decades pass in which records improve by no more than hundredths of a second. Take the 200-meter run: In 1968, the world record stood at 19.83 seconds; in 1996 Michael Johnson lowered it to 19.32 seconds — about a half a second in twenty-eight years.

None of this is good news for the human spirit. We need to keep desire alive. We depend on faith; records will fall only to those who believe it is possible. The heroes of my boyhood — Jim Ryun, Peter Snell, Herb Elliott, Steve Prefontaine, Billy Mills — achieved their status and success through sheer guts and work. They aspired to be gods — and to my high school cross-country mates and me, they *were* gods on some level. Yet the real reason we saw Pre and the others as heroes was that we secretly believed we were elementally equal. We were convinced that, if we only tried, if we did what they did, then we too could rank among the gods. To think that if they lived and ran today they would all be left in the dust by a herd of modern runners is devastating to my psyche. At our core we are endurance predators driven by dreams, spurred on by the antelope that we can't see but know is out there, somewhere, up ahead. To continue pushing, though, we must believe it's catchable — if only we apply ourselves.

Like the North American antelope's residual ability to outrun a cheetah — a cat that became extinct on the continent some 10,000 years ago — our abilities to run, throw, and jump are leftovers in our survival tool kits. As such, we use them in play because they are instinctually important to us. I'm not as athletically capable as an antelope or a bird or an Olympic athlete, but I enjoy my own capacities and I'm inspired to stretch them by seeing what others can do. I'm humbled by what is routine to the songbird or sandpiper, awed by their ability to fly unbelievably long distances to and from specific pinpoints on the globe.

Some might argue that, if I were a bird, I would not be able to enjoy my fantastic annual journeys, following the sun from perpetual daylight on the Arctic tundra to the pampas in Argentina and back again. But I think they are wrong. What makes the blackpoll warbler strike out south in the fall after a cold front is probably not fundamentally different from what motivates me to jog down a country road on a warm and sunny day. We're both responding to ancient urges. Proof that, in our case, it's impossible to extinguish our primal enthusiasm for the chase.

EDWARD HOAGLAND

Harpy Eagles

FROM *Orion*

HARPY EAGLES are the world's biggest birds of prey. Twice as heavy as golden eagles, they weigh nearly twenty pounds when adult, with a seven-foot wingspread, and stand three feet tall. They will nest a hundred feet up in a "spirit tree" — a jumbo silk-cotton, or kapok, emerging from the canopy — and lend a bristling supernal presence to the remaining Latin American rainforest. The harpy's stern visage and sinuous head plumage account for the English name, though *yaimo* is one of the many Indian names that various tribes have given them — this one by the Waiwais of the Brazil-Guyana border, who put eagle down in their hair to draw down the brightness and exuberance of the sun. Like other ultimate embodiments of wilderness large enough for people to take note of — mountain gorillas, Bengal tigers, polar bears, blue whales — harpy eagles aren't fenceable. Different whimsies or a different gyroscope and different feeding imperatives direct them. Their habitat is primeval but fragile, and their panache is more high-strung.

Harpies need great trees and a whole constellation of other life. Sloths, toucans, howler monkeys, and prismatic macaws to eat; broad-crotched limbs and lofty foliage to frequent; and not more than a few people lightly scattered about below, who don't shoot them as trophies, but value them alive as one spoke of the wheel of life. They're like a tuning fork with one pure pitch, not a contemporary multimedia medley composed of the chameleon habits and coloration of the mid-sized birds that manage to survive well now — blue herons that flap down and gobble a kitten off a Maryland lawn; cattle egrets picking bugs off a New York garbage barge; ravens stealing sandwiches from a snowmobile's seat pouch. Instead, they just materialize, a high-crested, black-beaked face peering down amid the highest leaves and boughs, or swerving to grab an anteater or an iguana from the trunk of a tree, then swinging upward secretively toward the nest so as to not betray its location by soaring in against the backdrop of the sky.

Spirit trees weren't cut by most indigenous tribes, and have an elegant look because of their airy architecture and pale, smooth bark. The trees harbor other intimations of the spirit-world besides the roosts and nests of these mysteriously severe-faced, huge, rare birds — which, sky-colored, care for and coach their single chick though an extraordinary cycle as long as thirty months. The Paratintin of the Amazon believed the harpy "Master of the Birds," who had painted everybody else's feathers; and the Tukano Indians of Colombia believed that harpies lived in the Milky Way.

Only a thousand years ago, there was an eagle in New Zealand that was at least half again as heavy as the harpy, and is thought to have hunted people. Its natural prey had been a man-tall, flightless bird called the moa by the ancestral Maoris, who when they first arrived on the two islands must have seemed to the eagles like an adjunct food, until the newcomers killed off both the moas and the eagles. We forget how much less safe and *explained* everything was then: sickness being a witchcraft delirium, history oral, sacredness in utter flux of forces.

Over most of the Earth, conservation is not a cause linked with the amenities of summer homes, camping trips, and other trappings of affluent living that our own bald eagles and golden eagles have begun to benefit from. In wildlife sanctuaries and on public lands here, people do want to see them, and ospreys, owls, and falcons, survive. Laws prohibit shooting them and the worst pesticides like DDT have been banned. Indeed, peregrines (once called "duck hawks") have been sufficiently adaptable that, instead of ducks, some of them are feeding on city pigeons, and nesting on skyscrapers and the girders of bridges, not their former habitat of wild-river cliffs and gorges. The countless locks and dams that clog our waterways and may play havoc with the fish, as stunned schools congregate in the turbid water underneath, have sometimes provided rallying points for bald eagles to feed abundantly and begin recovering their numbers. Ospreys, too — which also accept the convenience of man-made nesting platforms, where their traditional nesting trees have been leveled. Ospreys have even learned to dive for fish in decorative private ponds.

But affluence and therefore preservationist sympathies go with chilly climates, like the United States's and Europe's. The problem is, of course, that that's not where the biodiversity is. According to Scott Weidensaul's *Raptors: The Birds of Prey,* the lowland forests of the western Amazon alone harbor about as many species as all of Europe, plus the Middle East (thirty-nine). And the United States has around the same number. But Asia boasts ninety species, Africa eighty-eight, South America seventy-six. And these are the regions where creature-habitat is being obliterated at a pandemic pace, as people struggle for a standard of living of only modest comfort. All over, the largest or more complicated birds are fading, in favor of agile, versatile, less conspicuous types that can find something to eat almost anywhere, like our pileated woodpeckers (but not ivory-bills), or mute swans (but not the glorious trumpeters).

The harpy eagle — which the Kachuyana tribe thought was the most powerful sorcerer of the Otherworld; and the Yarabara people, "the Sun's warrior against the Forest Ogre" — has bear-sized claws and thick legs toughly scaled to fend off bites when it grabs a spider monkey, a kinkajou, or an agouti, though its grip is so drastic as to put most animals in a state of shock: they'll hang limp from a fist during the flight back to the nest. Sloths, which seem to be the harpy's most frequent prey, but are sensitive to habitat change, may have evolved their famous sluggardness as a defensive adaptation, making themselves hard for such a rapid hunter swerving by to see — a kind of motion-camouflage, even offering some protection against a dappled jaguar passing the sloth's tree trunk down below, but always prepared to climb. The sloth is additionally camouflaged because its hair is brown during the dry season but then turns green with algae growth at the onset of the rains.

The harpy is a "flying wolf," people say. But we like predators. Tigers, sharks, wolves fascinate us, like the spider in its treacherous orb, the *Just So* crocodile in the "great grey-green, greasy Limpopo River, all set about with fever trees," or a commando team that sneaks up on an enemy bunker and seizes hostages. Here in Vermont, I know plenty of men who only go into the woods during the fall, when they can carry a gun and shoot at living creatures. Otherwise there's no pleasure in being out-of-doors. And our empathy for eagles is further stirred because we like to get a tall perspective on many matters — corporate politics, family history — mounting up and sailing over the landscape of our lives. Hawklike, when young, we'll notice a comely member of the opposite sex and in our mind's eye want to swoop in, sink ourselves in them, feast slowly, and make their body ours. Or fly like eagles and swoop and seize a business deal, put our talons in it like a harpy grabbing a vegetarian sloth out of the gargantuan canopy, and carrying it off to eat in the spirit trees. But their ferocity is not gratuitous, like so much of ours. Even catching a capuchin monkey brachiating through the leaves, the ferocity is merely fleshly: a sparrowhawk's writ large.

Though tribal peoples sometimes raised fallen chicks as mascots or tethered them as village watchdogs, they didn't frivolously shoot such tree-crown citizens and spirits of the upper strata. The Cashinahua of the Peru-Brazil border, for instance, killed harpies only for ritual purposes — the headman's headdress and feather-backpack for ceremonies, or an initiation garb of white chest-down for children — and then only when they were in flight, not sitting or near the nest. The Heta tribe employed an eagle's wing to heal sick folk. But loggers coming in, and road-building crews, oil-field workers, gold prospectors with shotguns, rifles, and revolvers, and landless, slash-and-burn poor folk looking for new soil to farm, don't have such scruples and blast away, not just at this enormous, powerful bird, but at the howler monkeys, tegu lizards, and tree porcupines and anteaters they eat.

Not much survives the skinning and incineration of the forest. The constellation of bird cries, insect trills, trees rubbing against one another, and whispery twists of tangled wind, all had contributed to the tuning fork's perennial pitch, enigmatic, rich and sweet, except in storms. There was the anaconda, as green as swamp scum (for the scum may have been created as a cover for the anaconda); and the jaguar, marked like sunlight and leaf shadows (and the chief dry-ground spirit); and a hundred feet up a buttressed sacred tree trunk ten feet in diameter, the gray-blue harpy eagle, sky-colored but patched with black and white, like boughs and clouds as you'd look up. Under the crested plumage on its head, the white disklike face was sternly interrupted by the emphatic black triangle of its hooked beak. (The disk apparently captures sounds like an owl's facial disk for added acuity in hunting.)

A culture like the Waiwais' has regarded harpies as messengers from their own shamans up to the region of the sun, and a headdress of harpy feathers as equivalent in trading value to the life of a man. The Juruna, of central Brazil, believed mankind first captured Fire from an inattentive harpy. So when scientists climb to a nest in order to weigh and measure the single chick and collect bone fragments to identify the harpy diet, they find the adults flabbergasted at the intrusion, but not especially antagonistic or scared. Rather, the eagles seem to act like penguins in Antarctica, dumbfounded at the human apparition materializing from another planet. Unlike penguins, they've *seen* humans, but far below.

The occasional red-tailed hawk that sails or scuds over my field in Vermont, scouting for a mouse, shares the gleanings of that prey population with a pair of red foxes and also with a coyote family that includes it in their territory, not to mention a skunk or two, and the raccoons that den in a hollow maple tree inside the woods. Though the red-tails hunt in the woods a bit, as well as kiting, spiraling, and soaring over open country to catch sight of an unwary animal, a couple of hefty goshawks have set up shop in the conifers to veer and cruise through heavy cover after grouse or squirrels much faster than a red-tail could. Blue-backed and forest-oriented, goshawks dash like miniature harpies, with supple, palmate wings, hoping to surprise a snowshoe hare, or perhaps a blue jay preoccupied with teasing a woodpecker, a weasel engaged in consuming a red-backed vole, or a succulent mother opossum that has ventured unwisely out of her daytime shelter. Goshawks are surprisingly sizable for such good survivors in New England's woods, and get their living from a wide, flexing tail and big secondary feathers that deepen and shape their wings like hands for whirlwind turns and swift ambush strategies at close quarters.

Adaptation is the key, as abandoned farmlands either grow back to forest or become a suburban grid. Red-tailed hawks, for example, are replacing marsh hawks (harriers) over much of the remaining open marsh or meadowland because they are more ingeniously versatile, just as goshawks — secretive yet aggressive — have been replacing a less protean species, the Cooper's hawk, in many tracts of northern woods. During the winter, foraging in this new patchwork landscape, a goshawk or a red-tail may post itself in the red pines behind my birdfeeder. Goshawks, being faster, hunt songbirds from concealment year-round, whereas red-tails feed on a buffet of rodents, summer grasshoppers and garter snakes, stranded fish, and even roadkill. But when the snow gets deep, they too may need an avian meal and have learned that, flying at a flock of mourning doves pecking sunflower seeds on the ground, they can manage in the bumping panic to grab hold of the tardiest of these.

Crows, coons, coyotes, kestrels, and cowbirds are thriving because they compromise. They're wafflers; piggyback species. They're compressible. Whatever it takes, they say and do. A coyote that may have been only forty generations removed from the Midwestern plains reached Central Park in Manhattan last year, and cowbirds, which used to follow the buffalo herds by laying their eggs in other birds' nests and moving on, of course had already done the same. Kestrels (the old "sparrow hawk") — finding plenty of house sparrows, house wrens, and house finches — have never seemed more abundant, and turkey vultures, which profit, too, from disturbed habitat, are expanding their range. They will all still be our companions when harpies, tigers, and other hard-and-fast, kingpost creatures are gone.

I love crows and coyotes: sometimes I leave food for pairs of both at the edge of the woods in the winter, and once fed a coyote that had been hit by a car until it mended. "Brush wolves," they were called in frontier times (and "Trickster" by the Indians), so when the pristine forests were knocked down it hardly mattered to them. But the myths that tricksters engender are different. And when the biggest eagle or big cat dies out because it made no allowance for us, we too are diminished. A color is removed from the spectrum, a trumpet from the orchestra. Our retinas and eardrums deteriorate, and the world and our imaginations are flattened a little.

BILL JOY

Why the Future Doesn't Need Us

FROM *Wired*

FROM THE MOMENT I became involved in the creation of new technologies, their ethical dimensions have concerned me, but it was only in the autumn of 1998 that I became anxiously aware of how great are the dangers facing us in the twenty-first century. I can date the onset of my unease to the day I met Ray Kurzweil, the deservedly famous inventor of the first reading machine for the blind and many other amazing things.

Ray and I were both speakers at George Gilder's Telecosm conference, and I encountered him by chance in the bar of the hotel after both our sessions were over. I was sitting with John Searle, a Berkeley philosopher who studies consciousness. While we were talking, Ray approached and a conversation began, the subject of which haunts me to this day.

I had missed Ray's talk and the subsequent panel that Ray and John had been on, and they now picked right up where they'd left off, with Ray saying that the rate of improvement of technology was going to accelerate and that we were going to become robots or fuse with robots or something like that, and John countering that this couldn't happen, because the robots couldn't be conscious.

While I had heard such talk before, I had always felt sentient robots were in the realm of science fiction. But now, from someone I respected, I was hearing a strong argument that they were a near-term possibility. I was taken aback, especially given Ray's proven ability to imagine and create the future. I already knew that new technologies like genetic engineering and nanotechnology were giving us the power to remake the world, but a realistic and imminent scenario for intelligent robots surprised me.

It's easy to get jaded about such breakthroughs. We hear in the news almost every day of some kind of technological or scientific advance. Yet this was no ordinary prediction. In the hotel bar, Ray gave me a partial preprint of his then-forthcoming book *The Age of Spiritual Machines,* which outlined a utopia he foresaw — one in which humans gained near immortality by becoming one with robotic technology. On reading it, my sense of unease only intensified; I felt sure he had to be understating the dangers, understating the probability of a bad outcome along this path.

I found myself most troubled by a passage detailing a *dys*topian scenario:

The New Luddite Challenge

First let us postulate that the computer scientists succeed in developing intelligent machines that can do all things better than human beings can do them. In that case presumably all work will be done by vast, highly organized systems of machines and no human effort will be necessary. Either of two cases might occur. The machines might be permitted to make all of their own decisions without human oversight, or else human control over the machines might be retained.

If the machines are permitted to make all their own decisions, we can't make any conjectures as to the results, because it is impossible to guess how such machines might behave. We only point out that the fate of the human race would be at the mercy of the machines. It might be argued that the human race would never be foolish enough to hand over all the power to the machines. But we are suggesting neither that the human race would voluntarily turn power over to the machines nor that the machines would willfully seize power. What we do suggest is that the human race might easily permit itself to drift into a position of such dependence on the machines that it would have no practical choice but to accept all of the machines' decisions. As society and the problems that face it become more and more complex and machines become more and more intelligent, people will let machines make more of their decisions for them, simply because machine-made decisions will bring better results than man-made ones. Eventually a stage may be reached at which the decisions necessary to keep the system running will be so complex that human beings will be incapable of making them intelligently. At that stage the machines will be in effective control. People won't be able to just turn the machines off, because they will be so dependent on them that turning them off would amount to suicide.

On the other hand it is possible that human control over the ma-

> chines may be retained. In that case the average man may have control over certain private machines of his own, such as his car or his personal computer, but control over large systems of machines will be in the hands of a tiny elite — just as it is today, but with two differences. Due to improved techniques the elite will have greater control over the masses; and because human work will no longer be necessary the masses will be superfluous, a useless burden on the system. If the elite is ruthless they may simply decide to exterminate the mass of humanity. If they are humane they may use propaganda or other psychological or biological techniques to reduce the birth rate until the mass of humanity becomes extinct, leaving the world to the elite. Or, if the elite consists of soft-hearted liberals, they may decide to play the role of good shepherds to the rest of the human race. They will see to it that everyone's physical needs are satisfied, that all children are raised under psychologically hygienic conditions, that everyone has a wholesome hobby to keep him busy, and that anyone who may become dissatisfied undergoes "treatment" to cure his "problem." Of course, life will be so purposeless that people will have to be biologically or psychologically engineered either to remove their need for the power process or make them "sublimate" their drive for power into some harmless hobby. These engineered human beings may be happy in such a society, but they will most certainly not be free. They will have been reduced to the status of domestic animals.[1]

In the book, you don't discover until you turn the page that the author of this passage is Theodore Kaczynski — the Unabomber. I am no apologist for Kaczynski. His bombs killed three people during a seventeen-year terror campaign and wounded many others. One of his bombs gravely injured my friend David Gelernter, one of the most brilliant and visionary computer scientists of our time. Like many of my colleagues, I felt that I could easily have been the Unabomber's next target.

Kaczynski's actions were murderous and, in my view, criminally insane. He is clearly a Luddite, but simply saying this does not dismiss his argument; as difficult as it is for me to acknowledge, I saw some merit in the reasoning in this single passage. I felt compelled to confront it.

Kaczynski's dystopian vision describes unintended consequences, a well-known problem with the design and use of technology, and one that is clearly related to Murphy's law — "Anything that can go wrong, will." (Actually, this is Finagle's law, which in itself shows that Finagle was right.) Our overuse of antibiotics has led to what may be the biggest such problem so far: the emergence of antibiotic-resistant and much more dangerous bacteria. Similar things happened when attempts to eliminate malarial mosquitoes using DDT caused them to acquire DDT resistance; malarial parasites likewise acquired multidrug-resistant genes.[2]

The cause of many such surprises seems clear: The systems involved are complex, involving interaction among and feedback between many parts. Any changes to such a system will cascade in ways that are difficult to predict; this is especially true when human actions are involved.

I started showing friends the Kaczynski quote from *The Age of Spiritual Machines;* I would hand them Kurzweil's book, let them read the quote, and then watch their reaction as they discovered who had written it. At around the same time, I found Hans Moravec's book *Robot: Mere Machine to Transcendent Mind.* Moravec is one of the leaders in robotics research, and was a founder of the world's largest robotics research program, at Carnegie Mellon University. *Robot* gave me more material to try out on my friends — material surprisingly supportive of Kaczynski's argument. For example:

The Short Run (Early 2000s)

> Biological species almost never survive encounters with superior competitors. Ten million years ago, South and North America were separated by a sunken Panama isthmus. South America, like Australia today, was populated by marsupial mammals, including pouched equivalents of rats, deers, and tigers. When the isthmus connecting North and South America rose, it took only a few thousand years for the northern placental species, with slightly more effective metabolisms and reproductive and nervous systems, to displace and eliminate almost all the southern marsupials.
>
> In a completely free marketplace, superior robots would surely affect humans as North American placentals affected South American marsupials (and as humans have affected countless species). Robotic industries would compete vigorously among themselves for matter, energy, and space, incidentally driving their price beyond human reach. Unable to afford the necessities of life, biological humans would be squeezed out of existence.
>
> There is probably some breathing room, because we do not live in a

> completely free marketplace. Government coerces nonmarket behavior, especially by collecting taxes. Judiciously applied, governmental coercion could support human populations in high style on the fruits of robot labor, perhaps for a long while.

A textbook dystopia — and Moravec is just getting wound up. He goes on to discuss how our main job in the twenty-first century will be "ensuring continued cooperation from the robot industries" by passing laws decreeing that they be "nice,"[3] and to describe how seriously dangerous a human can be "once transformed into an unbounded superintelligent robot." Moravec's view is that the robots will eventually succeed us — that humans clearly face extinction.

I decided it was time to talk to my friend Danny Hillis. Danny became famous as the cofounder of Thinking Machines Corporation, which built a very powerful parallel supercomputer. Despite my current job title of Chief Scientist at Sun Microsystems, I am more a computer architect than a scientist, and I respect Danny's knowledge of the information and physical sciences more than that of any other single person I know. Danny is also a highly regarded futurist who thinks long-term — four years ago he started the Long Now Foundation, which is building a clock designed to last 10,000 years, in an attempt to draw attention to the pitifully short attention span of our society.

So I flew to Los Angeles for the express purpose of having dinner with Danny and his wife, Pati. I went through my now-familiar routine, trotting out the ideas and passages that I found so disturbing. Danny's answer — directed specifically at Kurzweil's scenario of humans merging with robots — came swiftly, and quite surprised me. He said, simply, that the changes would come gradually, and that we would get used to them.

But I guess I wasn't totally surprised. I had seen a quote from Danny in Kurzweil's book in which he said, "I'm as fond of my body as anyone, but if I can be two hundred with a body of silicon, I'll take it." It seemed that he was at peace with this process and its attendant risks, while I was not.

While talking and thinking about Kurzweil, Kaczynski, and Moravec, I suddenly remembered a novel I had read almost twenty years ago — *The White Plague,* by Frank Herbert — in which a molecular biologist is driven insane by the senseless murder of his family. To seek revenge he constructs and disseminates a new and highly contagious plague that kills widely but selectively. (We're lucky Kaczynski was a mathematician, not a molecular biologist.) I was also reminded of the Borg of *Star Trek*, a hive of partly biological, partly robotic creatures with a strong destructive streak. Borg-like disasters are a staple of science fiction, so why hadn't I been more concerned about such robotic dystopias earlier? Why weren't other people more concerned about these nightmarish scenarios?

Part of the answer certainly lies in our attitude toward the new — in our bias toward instant familiarity and unquestioning acceptance. Accustomed to living with almost routine scientific breakthroughs, we have yet to come to terms with the fact that the most compelling twenty-first-century technologies — robotics, genetic engineering, and nanotechnology — pose a different threat than the technologies that have come before. Specifically, robots, engineered organisms, and nanobots share a dangerous amplifying factor: They can self-replicate. A bomb is blown up only once — but one bot can become many, and quickly get out of control.

Much of my work over the past twenty-five years has been on computer networking, where the sending and receiving of messages creates the opportunity for out-of-control replication. But while replication in a computer or a computer network can be a nuisance, at worst it disables a machine or takes down a network or network service. Uncontrolled self-replication in these newer technologies runs a much greater risk: a risk of substantial damage in the physical world.

Each of these technologies also offers untold promise: The vision of near immortality that Kurzweil sees in his robot dreams drives us forward; genetic engineering may soon provide treatments, if not outright cures, for most diseases; and nanotechnology and nanomedicine can address yet more ills. Together they could significantly extend our average life span and improve the quality of our lives. Yet, with each of these technologies, a sequence of small, individually sensible advances leads to an accumulation of great power and, concomitantly, great danger.

What was different in the twentieth century? Certainly, the technologies underlying the weapons of mass destruction (WMD) — nuclear, biological, and chemical (NBC) — were powerful, and the weapons an enormous threat. But building nuclear weapons required, at least for a time, access to both rare — indeed, effectively unavailable — raw materials and highly protected information; biological and chemical weapons programs also tended to require large-scale activities.

The twenty-first-century technologies — genetics, nanotechnology, and robotics (GNR) — are so powerful that they can spawn whole new classes of accidents and abuses. Most dangerously, for the first time, these accidents and abuses are widely within the reach of individuals or small groups. They will not require large facilities or rare raw materials. Knowledge alone will enable the use of them.

Thus we have the possibility not just of weapons of mass destruction but of knowledge-enabled mass destruction (KMD), this destructiveness hugely amplified by the power of self-replication.

I think it is no exaggeration to say we are on the cusp of the further perfection of extreme evil, an evil whose possibility spreads well beyond that which weapons of mass destruction bequeathed to the nation-states, on to a surprising and terrible empowerment of extreme individuals.

Nothing about the way I got involved with computers suggested to me that I was going to be facing these kinds of issues. My life has been driven by a deep need to ask questions and find answers. When I was three, I was already reading, so my father took me to the elementary school, where I sat on the principal's lap and read him a story. I started school early, later skipped a grade, and escaped into books — I was incredibly motivated to learn. I asked lots of questions, often driving adults to distraction.

As a teenager I was very interested in science and technology. I wanted to be a ham radio operator but didn't have the money to buy the equipment. Ham radio was the Internet of its time: very addictive, and quite solitary. Money issues aside, my mother put her foot down — I was not to be a ham; I was antisocial enough already.

I may not have had many close friends, but I was awash in ideas. By high school, I had discovered the great science fiction writers. I remember especially Heinlein's *Have Spacesuit Will Travel* and Asimov's *I, Robot,* with its Three Laws of Robotics. I was enchanted by the descriptions of space travel, and wanted to have a telescope to look at the stars; since I had no money to buy or make one, I checked books on telescope-making out of the library and read about making them instead. I soared in my imagination.

Thursday nights my parents went bowling, and we kids stayed home alone. It was the night of Gene Roddenberry's original *Star Trek,* and the program made a big impression on me. I came to accept its notion that humans had a future in space, Western-style, with big heroes and adventures. Roddenberry's vision of the centuries to come was one with strong moral values, embodied in codes like the Prime Directive: to not interfere in the development of less technologically advanced civilizations. This had an incredible appeal to me; ethical humans, not robots, dominated this future, and I took Roddenberry's dream as part of my own.

I excelled in mathematics in high school, and when I went to the University of Michigan as an undergraduate engineering student I took the advanced curriculum of the mathematics majors. Solving math problems was an exciting challenge, but when I discovered computers I found something much more interesting: a machine into which you could put a program that attempted to solve a problem, after which the machine quickly checked the solution. The computer had a clear notion of correct and incorrect, true and false. Were my ideas correct? The machine could tell me. This was very seductive.

I was lucky enough to get a job programming early supercomputers and discovered the amazing power of large machines to numerically simulate advanced designs. When I went to graduate school at UC Berkeley in the mid-1970s, I started staying up late, often all night, inventing new worlds inside the machines. Solving problems. Writing the code that argued so strongly to be written.

In *The Agony and the Ecstasy,* Irving Stone's biographical novel of Michelangelo, Stone described vividly how Michelangelo released the statues from the stone, "breaking the marble spell," carving from the images in his mind.[4] In my most ecstatic moments, the software in the computer emerged in the same way. Once I had imagined it in my mind I felt that it was already there in the machine, waiting to be released. Staying up all night seemed a small price to pay to free it — to give the ideas concrete form.

After a few years at Berkeley I started to send out some of the software I had written — an instructional Pascal system, Unix utilities, and a text editor called vi (which is still, to my surprise, widely used more than twenty years later) — to others who had similar small PDP-11 and VAX minicomputers. These adventures in software eventually turned into the Berkeley version of the Unix operating system, which became a personal "success disaster" — so many people wanted it that I never finished my Ph.D. Instead I got a job working for Darpa putting Berkeley Unix on the Internet and fixing it to be reliable and to run large research applications well. This was all great fun and very rewarding. And, frankly, I saw no robots here, or anywhere near.

Still, by the early 1980s, I was drowning. The Unix releases were very successful, and my little project of one soon had money and some staff, but the problem at Berkeley was always office space rather than money — there wasn't room for the help the project needed, so when the other founders of Sun Microsystems showed up I jumped at the chance to join them. At Sun, the long hours continued into the early days of workstations and personal computers, and I have enjoyed participating in the creation of advanced microprocessor technologies and Internet technologies such as Java and Jini.

From all this, I trust it is clear that I am not a Luddite. I have always, rather, had a strong belief in the value of the scientific search for truth and in the ability of great engineering to bring material progress. The Industrial Revolution has immeasurably improved everyone's life over the last couple hundred years, and I always expected my career to involve the building of worthwhile solutions to real problems, one problem at a time.

I have not been disappointed. My work has had more impact than I had ever hoped for and has been more widely used than I could have reasonably expected. I have spent the last twenty years still trying to figure out how to make computers as reliable as I want them to be (they are not nearly there yet) and how to make them simple to use (a goal that has met with even less relative success). Despite some progress, the problems that remain seem even more daunting.

But while I was aware of the moral dilemmas surrounding technology's consequences in fields like weapons research, I did not expect that I would confront such issues in my own field, or at least not so soon.

*

Perhaps it is always hard to see the bigger impact while you are in the vortex of a change. Failing to understand the consequences of our inventions while we are in the rapture of discovery and innovation seems to be a common fault of scientists and technologists; we have long been driven by the overarching desire to know that is the nature of science's quest, not stopping to notice that the progress to newer and more powerful technologies can take on a life of its own.

I have long realized that the big advances in information technology come not from the work of computer scientists, computer architects, or electrical engineers, but from that of physical scientists. The physicists Stephen Wolfram and Brosl Hasslacher introduced me, in the early 1980s, to chaos theory and nonlinear systems. In the 1990s, I learned about complex systems from conversations with Danny Hillis, the biologist Stuart Kauffman, the Nobel-laureate physicist Murray Gell-Mann, and others. Most recently, Hasslacher and the electrical engineer and device physicist Mark Reed have been giving me insight into the incredible possibilities of molecular electronics.

In my own work, as codesigner of three microprocessor architectures — SPARC, picoJava, and MAJC — and as the designer of several implementations thereof, I've been afforded a deep and firsthand acquaintance with Moore's law. For decades, Moore's law has correctly predicted the exponential rate of improvement of semiconductor technology. Until last year I believed that the rate of advances predicted by Moore's law might continue only until roughly 2010, when some physical limits would begin to be reached. It was not obvious to me that a new technology would arrive in time to keep performance advancing smoothly.

But because of the recent rapid and radical progress in molecular electronics — where individual atoms and molecules replace lithographically drawn transistors — and related nanoscale technologies, we should be able to meet or exceed the Moore's law rate of progress for another thirty years. By 2030, we are likely to be able to build machines, in quantity, a million times as powerful as the personal computers of today — sufficient to implement the dreams of Kurzweil and Moravec.

As this enormous computing power is combined with the manipulative advances of the physical sciences and the new, deep understandings in genetics, enormous transformative power is being unleashed. These combinations open up the opportunity to completely redesign the world, for better or worse: The replicating and evolving processes that have been confined to the natural world are about to become realms of human endeavor.

In designing software and microprocessors, I have never had the feeling that I was designing an intelligent machine. The software and hardware is so fragile and the capabilities of the machine to "think" so clearly absent that, even as a possibility, this has always seemed very far in the future.

But now, with the prospect of human-level computing power in about thirty years, a new idea suggests itself: that I may be working to create tools that will enable the construction of the technology that may replace our species. How do I feel about this? Very uncomfortable. Having struggled my entire career to build reliable software systems, it seems to me more than likely that this future will not work out as well as some people may imagine. My personal experience suggests we tend to overestimate our design abilities.

Given the incredible power of these new technologies, shouldn't we be asking how we can best coexist with them? And if our own extinction is a likely, or even possible, outcome of our technological development, shouldn't we proceed with great caution?

The dream of robotics is, first, that intelligent machines can do our work for us, allowing us lives of leisure, restoring us to Eden. Yet in his history of such ideas, *Darwin Among the Machines,* George Dyson warns: "In the game of life and evolution there are three players at the table: human beings, nature, and machines. I am firmly on the side of nature. But nature, I suspect, is on the side of the machines." As we have seen, Moravec agrees, believing we may well not survive the encounter with the superior robot species.

How soon could such an intelligent robot be built? The coming advances in computing power seem to make it possible by 2030. And once an intelligent robot exists, it is only a small step to a robot species — to an intelligent robot that can make evolved copies of itself.

A second dream of robotics is that we will gradually replace ourselves with our robotic technology, achieving near immortality by downloading our consciousnesses; it is this process that Danny Hillis thinks we will gradually get used to and that Ray Kurzweil elegantly details in *The Age of Spiritual Machines.* (We are beginning to see intimations of this in the implantation of computer devices into the human body.)

But if we are downloaded into our technology, what are the chances that we will thereafter be ourselves or even human? It seems to me far more likely that a robotic existence would not be like a human one in any sense that we understand, that the robots would in no sense be our children, that on this path our humanity may well be lost.

Genetic engineering promises to revolutionize agriculture by increasing crop yields while reducing the use of pesticides; to create tens of thousands of novel species of bacteria, plants, viruses, and animals; to replace reproduction, or supplement it, with cloning; to create cures for many diseases, increasing our life span and our quality of life; and much, much more. We now know with certainty that these profound changes in the biological sciences are imminent and will challenge all our notions of what life is.

Technologies such as human cloning have in particular raised our awareness of the profound ethical and moral issues we face. If, for example, we were to reengineer ourselves into several separate and unequal species using the power of genetic engineering, then we would threaten the notion of equality that is the very cornerstone of our democracy.

Given the incredible power of genetic engineering, it's no surprise that there are significant safety issues in its use. My friend Amory Lovins recently cowrote, along with Hunter Lovins, an editorial that provides an ecological view of some of these dangers. Among their concerns: that "the new botany aligns the development of plants with their economic, not evolutionary, success." Amory's long career has been focused on energy and resource efficiency by taking a whole-system view of human-made systems; such a whole-system view often finds simple, smart solutions to otherwise seemingly difficult problems, and is usefully applied here as well.

After reading the Lovins' editorial, I saw an op-ed by Gregg Easterbrook in the *New York Times* (November 19, 1999) about genetically engineered crops, under the headline: "Food for the Future: Someday, rice will have built-in vitamin A. Unless the Luddites win."

Are Amory and Hunter Lovins Luddites? Certainly not. I believe we all would agree that golden rice, with its built-in vitamin A, is probably a good thing, if developed with proper care and respect for the likely dangers in moving genes across species boundaries.

Awareness of the dangers inherent in genetic engineering is beginning to grow, as reflected in the Lovins' editorial. The general public is aware of, and uneasy about, genetically modified foods, and seems to be rejecting the notion that such foods should be permitted to be unlabeled.

But genetic engineering technology is already very far along. As the Lovins note, the USDA has already approved about fifty genetically engineered crops for unlimited release; more than half of the world's soybeans and a third of its corn now contain genes spliced in from other forms of life.

While there are many important issues here, my own major concern with genetic engineering is narrower: that it gives the power — whether militarily, accidentally, or in a deliberate terrorist act — to create a White Plague.

The many wonders of nanotechnology were first imagined by the Nobel-laureate physicist Richard Feynman in a speech he gave in 1959, subsequently published under the title "There's Plenty of Room at the Bottom." The book that made a big impression on me, in the mid-'80s, was Eric Drexler's *Engines of Creation,* in which he described beautifully how manipulation of matter at the atomic level could create a utopian future of abundance, where just about everything could be made cheaply, and almost any imaginable disease or physical problem could be solved using nanotechnology and artificial intelligences.

A subsequent book, *Unbounding the Future: The Nanotechnology Revolution,* which Drexler cowrote, imagines some of the changes that might take place in a world where we had molecular-level "assemblers." Assemblers could make possible incredibly low-cost solar power, cures for cancer and the common cold by augmentation of the human immune system, essentially complete cleanup of the environment, incredibly inexpensive pocket supercomputers — in fact, any product would be manufacturable by assemblers at a cost no greater than that of wood — spaceflight more accessible than transoceanic travel today, and restoration of extinct species.

I remember feeling good about nanotechnology after reading *Engines of Creation.* As a technologist, it gave me a sense of calm —

that is, nanotechnology showed us that incredible progress was possible, and indeed perhaps inevitable. If nanotechnology was our future, then I didn't feel pressed to solve so many problems in the present. I would get to Drexler's utopian future in due time; I might as well enjoy life more in the here and now. It didn't make sense, given his vision, to stay up all night, all the time.

Drexler's vision also led to a lot of good fun. I would occasionally get to describe the wonders of nanotechnology to others who had not heard of it. After teasing them with all the things Drexler described I would give a homework assignment of my own: "Use nanotechnology to create a vampire; for extra credit create an antidote."

With these wonders came clear dangers, of which I was acutely aware. As I said at a nanotechnology conference in 1989, "We can't simply do our science and not worry about these ethical issues."[5] But my subsequent conversations with physicists convinced me that nanotechnology might not even work — or, at least, it wouldn't work anytime soon. Shortly thereafter I moved to Colorado, to a skunk works I had set up, and the focus of my work shifted to software for the Internet, specifically on ideas that became Java and Jini.

Then, last summer, Brosl Hasslacher told me that nanoscale molecular electronics was now practical. This was *new* news, at least to me, and I think to many people — and it radically changed my opinion about nanotechnology. It sent me back to *Engines of Creation.* Rereading Drexler's work after more than ten years, I was dismayed to realize how little I had remembered of its lengthy section called "Dangers and Hopes," including a discussion of how nanotechnologies can become "engines of destruction." Indeed, in my rereading of this cautionary material today, I am struck by how naive some of Drexler's safeguard proposals seem, and how much greater I judge the dangers to be now than even he seemed to then. (Having anticipated and described many technical and political problems with nanotechnology, Drexler started the Foresight Institute in the late 1980s "to help prepare society for anticipated advanced technologies" — most important, nanotechnology.)

The enabling breakthrough to assemblers seems quite likely within the next twenty years. Molecular electronics — the new subfield of nanotechnology where individual molecules are circuit elements — should mature quickly and become enormously lucrative within this decade, causing a large incremental investment in all nanotechnologies.

Unfortunately, as with nuclear technology, it is far easier to create destructive uses for nanotechnology than constructive ones. Nanotechnology has clear military and terrorist uses, and you need not be suicidal to release a massively destructive nanotechnological device — such devices can be built to be selectively destructive, affecting, for example, only a certain geographical area or a group of people who are genetically distinct.

An immediate consequence of the Faustian bargain in obtaining the great power of nanotechnology is that we run a grave risk — the risk that we might destroy the biosphere on which all life depends.

As Drexler explained:

> "Plants" with "leaves" no more efficient than today's solar cells could out-compete real plants, crowding the biosphere with an inedible foliage. Tough omnivorous "bacteria" could out-compete real bacteria: They could spread like blowing pollen, replicate swiftly, and reduce the biosphere to dust in a matter of days. Dangerous replicators could easily be too tough, small, and rapidly spreading to stop — at least if we make no preparation. We have trouble enough controlling viruses and fruit flies.
>
> Among the cognoscenti of nanotechnology, this threat has become known as the "gray goo problem." Though masses of uncontrolled replicators need not be gray or gooey, the term "gray goo" emphasizes that replicators able to obliterate life might be less inspiring than a single species of crabgrass. They might be superior in an evolutionary sense, but this need not make them valuable.
>
> The gray goo threat makes one thing perfectly clear: We cannot afford certain kinds of accidents with replicating assemblers.

Gray goo would surely be a depressing ending to our human adventure on Earth, far worse than mere fire or ice, and one that could stem from a simple laboratory accident.[6] Oops.

It is most of all the power of destructive self-replication in genetics, nanotechnology, and robotics (GNR) that should give us pause. Self-replication is the modus operandi of genetic engineering, which uses the machinery of the cell to replicate its designs, and the prime danger underlying gray goo in nanotechnology. Stories of run-amok robots like the Borg, replicating or mutating to escape from the ethical constraints imposed on them by their creators, are well established in our science fiction books and movies. It is even possible that self-replication may be more fundamental than we thought, and hence harder — or even impossible — to control. A recent article by Stuart Kauffman in *Nature* titled "Self-Replication: Even Peptides Do It" discusses the discovery that a 32-amino-acid peptide can "autocatalyse its own synthesis." We don't know how widespread this ability is, but Kauffman notes that it may hint at "a route to self-reproducing molecular systems on a basis far wider than Watson-Crick base-pairing."[7]

In truth, we have had in hand for years clear warnings of the dangers inherent in widespread knowledge of GNR technologies — of the possibility of knowledge alone enabling mass destruction. But these warnings haven't been widely publicized; the public discussions have been clearly inadequate. There is no profit in publicizing the dangers.

The nuclear, biological, and chemical (NBC) technologies used in twentieth-century weapons of mass destruction were and are largely military, developed in government laboratories. In sharp contrast, the twenty-first-century GNR technologies have clear commercial uses and are being developed almost exclusively by corporate enterprises. In this age of triumphant commercialism, technology — with science as its handmaiden — is delivering a series of almost magical inventions that are the most phenomenally lucrative ever seen. We are aggressively pursuing the promises of these new technologies within the now-unchallenged system of global capitalism and its manifold financial incentives and competitive pressures.

> This is the first moment in the history of our planet when any species, by its own voluntary actions, has become a danger to itself — as well as to vast numbers of others. . . .
>
> It might be a familiar progression, transpiring on many worlds — a planet, newly formed, placidly revolves around its star; life slowly forms; a kaleidoscopic procession of creatures evolves; intelligence emerges which, at least up to a point, confers enormous survival value; and then technology is invented. It dawns on them that there are such things as laws of Nature, that these laws can be revealed by experiment, and that

knowledge of these laws can be made both to save and to take lives, both on unprecedented scales. Science, they recognize, grants immense powers. In a flash, they create world-altering contrivances. Some planetary civilizations see their way through, place limits on what may and what must not be done, and safely pass through the time of perils. Others, not so lucky or so prudent, perish.

That is Carl Sagan, writing in 1994, in *Pale Blue Dot,* a book describing his vision of the human future in space. I am only now realizing how deep his insight was, and how sorely I miss, and will miss, his voice. For all its eloquence, Sagan's contribution was not least that of simple common sense — an attribute that, along with humility, many of the leading advocates of the twenty-first-century technologies seem to lack.

I remember from my childhood that my grandmother was strongly against the overuse of antibiotics. She had worked since before the First World War as a nurse and had a commonsense attitude that taking antibiotics, unless they were absolutely necessary, was bad for you.

It is not that she was an enemy of progress. She saw much progress in an almost seventy-year nursing career; my grandfather, a diabetic, benefited greatly from the improved treatments that became available in his lifetime. But she, like many levelheaded people, would probably think it greatly arrogant for us, now, to be designing a robotic "replacement species," when we obviously have so much trouble making relatively simple things work, and so much trouble managing — or even understanding — ourselves.

I realize now that she had an awareness of the nature of the order of life, and of the necessity of living with and respecting that order. With this respect comes a necessary humility that we, with our early-twenty-first-century chutzpah, lack at our peril. The commonsense view, grounded in this respect, is often right, in advance of the scientific evidence. The clear fragility and inefficiencies of the human-made systems we have built should give us all pause; the fragility of the systems I have worked on certainly humbles me.

We should have learned a lesson from the making of the first atomic bomb and the resulting arms race. We didn't do well then, and the parallels to our current situation are troubling.

The effort to build the first atomic bomb was led by the brilliant physicist J. Robert Oppenheimer. Oppenheimer was not naturally interested in politics but became painfully aware of what he perceived as the grave threat to Western civilization from the Third Reich, a threat surely grave because of the possibility that Hitler might obtain nuclear weapons. Energized by this concern, he brought his strong intellect, passion for physics, and charismatic leadership skills to Los Alamos and led a rapid and successful effort by an incredible collection of great minds to quickly invent the bomb.

What is striking is how this effort continued so naturally after the initial impetus was removed. In a meeting shortly after V-E Day with some physicists who felt that perhaps the effort should stop, Oppenheimer argued to continue. His stated reason seems a bit strange: not because of the fear of large casualties from an invasion of Japan, but because the United Nations, which was soon to be formed, should have foreknowledge of atomic weapons. A more likely reason the project continued is the momentum that had built up — the first atomic test, Trinity, was nearly at hand.

We know that in preparing this first atomic test the physicists proceeded despite a large number of possible dangers. They were initially worried, based on a calculation by Edward Teller, that an atomic explosion might set fire to the atmosphere. A revised calculation reduced the danger of destroying the world to a three-in-a-million chance. (Teller says he was later able to dismiss the prospect of atmospheric ignition entirely.) Oppenheimer, though, was sufficiently concerned about the result of Trinity that he arranged for a possible evacuation of the southwest part of the state of New Mexico. And, of course, there was the clear danger of starting a nuclear arms race.

Within a month of that first, successful test, two atomic bombs destroyed Hiroshima and Nagasaki. Some scientists had suggested that the bomb simply be demonstrated, rather than dropped on Japanese cities — saying that this would greatly improve the chances for arms control after the war — but to no avail. With the tragedy of Pearl Harbor still fresh in Americans' minds, it would have been very difficult for President Truman to order a demonstration of the weapons rather than use them as he did — the desire to quickly end the war and save the lives that would have been lost in any invasion of Japan was very strong. Yet the overriding truth was probably very simple: As the physicist Freeman Dyson later said, "The reason that it was dropped was just that nobody had the courage or the foresight to say no."

It's important to realize how shocked the physicists were in the aftermath of the bombing of Hiroshima, on August 6, 1945. They describe a series of waves of emotion: first, a sense of fulfillment that the bomb worked, then horror at all the people that had been killed, and then a convincing feeling that on no account should another bomb be dropped. Yet of course another bomb was dropped, on Nagasaki, only three days after the bombing of Hiroshima.

In November 1945, three months after the atomic bombings, Oppenheimer stood firmly behind the scientific attitude, saying, "It is not possible to be a scientist unless you believe that the knowledge of the world, and the power which this gives, is a thing which is of intrinsic value to humanity, and that you are using it to help in the spread of knowledge and are willing to take the consequences."

Oppenheimer went on to work, with others, on the Acheson-Lilienthal report, which, as Richard Rhodes says in his recent book *Visions of Technology*, "found a way to prevent a clandestine nuclear arms race without resorting to armed world government"; their suggestion was a form of relinquishment of nuclear weapons work by nation-states to an international agency.

This proposal led to the Baruch Plan, which was submitted to the United Nations in June 1946 but never adopted (perhaps because, as Rhodes suggests, Bernard Baruch had "insisted on burdening the plan with conventional sanctions," thereby inevitably dooming it, even though it would "almost certainly have been rejected by Stalinist Russia anyway"). Other efforts to promote sensible steps toward internationalizing nuclear power to prevent an arms race ran afoul either of U.S. politics and internal distrust, or distrust by the Soviets. The opportunity to avoid the arms race was lost, and very quickly.

Two years later, in 1948, Oppenheimer seemed to have reached another stage in his thinking, saying, "In some sort of crude sense which no vulgarity, no humor, no overstatement can quite extinguish, the physicists have known sin; and this is a knowledge they cannot lose."

In 1949, the Soviets exploded an atom bomb. By 1955, both the United States and the Soviet Union had tested hydrogen bombs suitable for delivery by aircraft. And so the nuclear arms race began.

Nearly twenty years ago, in the documentary *The Day After Trinity*, Freeman Dyson summarized the scientific attitudes that brought us to the nuclear precipice:

> I have felt it myself. The glitter of nuclear weapons. It is irresistible if you come to them as a scientist. To feel it's there in your hands, to release this energy that fuels the stars, to let it do your bidding. To perform these miracles, to lift a million tons of rock into the sky. It is something that gives people an illusion of illimitable power, and it is, in some ways, responsible for all our troubles — this, what you might call technical arrogance, that overcomes people when they see what they can do with their minds.[8]

Now, as then, we are creators of new technologies and stars of the imagined future, driven — this time by great financial rewards and global competition — despite the clear dangers, hardly evaluating what it may be like to try to live in a world that is the realistic outcome of what we are creating and imagining.

In 1947, *The Bulletin of the Atomic Scientists* began putting a Doomsday Clock on its cover. For more than fifty years, it has shown an estimate of the relative nuclear danger we have faced, reflecting the changing international conditions. The hands on the clock have moved fifteen times and today, standing at nine minutes to midnight, reflect continuing and real danger from nuclear weapons. The recent addition of India and Pakistan to the list of nuclear powers has increased the threat of failure of the nonproliferation goal, and this danger was reflected by moving the hands closer to midnight in 1998.

In our time, how much danger do we face, not just from nuclear weapons, but from all of these technologies? How high are the extinction risks?

The philosopher John Leslie has studied this question and concluded that the risk of human extinction is at least thirty percent,[9] while Ray Kurzweil believes we have "a better than even chance of making it through," with the caveat that he has "always been accused of being an optimist." Not only are these estimates not encouraging, but they do not include the probability of many horrid outcomes that lie short of extinction.

Faced with such assessments, some serious people are already suggesting that we simply move beyond Earth as quickly as possible. We would colonize the galaxy using von Neumann probes, which hop from star system to star system, replicating as they go. This step will almost certainly be necessary 5 billion years from now (or sooner if our solar system is disastrously impacted by the impending collision of our galaxy with the Andromeda galaxy within the next 3 billion years), but if we take Kurzweil and Moravec at their word it might be necessary by the middle of this century.

What are the moral implications here? If we must move beyond Earth this quickly in order for the species to survive, who accepts the responsibility for the fate of those (most of us, after all) who are left behind? And even if we scatter to the stars, isn't it likely that we may take our problems with us or find, later, that they have followed us? The fate of our species on Earth and our fate in the galaxy seem inextricably linked.

Another idea is to erect a series of shields to defend against each of the dangerous technologies. The Strategic Defense Initiative, proposed by the Reagan administration, was an attempt to design such a shield against the threat of a nuclear attack from the Soviet Union. But as Arthur C. Clarke, who was privy to discussions about the project, observed: "Though it might be possible, at vast expense, to construct local defense systems that would 'only' let through a few percent of ballistic missiles, the much touted idea of a national umbrella was nonsense. Luis Alvarez, perhaps the greatest experimental physicist of this century, remarked to me that the advocates of such schemes were 'very bright guys with no common sense.'"

Clarke continued: "Looking into my often cloudy crystal ball, I suspect that a total defense might indeed be possible in a century or so. But the technology involved would produce, as a by-product, weapons so terrible that no one would bother with anything as primitive as ballistic missiles."[10]

In *Engines of Creation*, Eric Drexler proposed that we build an active nanotechnological shield — a form of immune system for the biosphere — to defend against dangerous replicators of all kinds that might escape from laboratories or otherwise be maliciously created. But the shield he proposed would itself be extremely dangerous — nothing could prevent it from developing autoimmune problems and attacking the biosphere itself.[11]

Similar difficulties apply to the construction of shields against robotics and genetic engineering. These technologies are too powerful to be shielded against in the time frame of interest; even if it were possible to implement defensive shields, the side effects of their development would be at least as dangerous as the technologies we are trying to protect against.

These possibilities are all thus either undesirable or unachievable or both. The only realistic alternative I see is relinquishment: to limit development of the technologies that are too dangerous, by limiting our pursuit of certain kinds of knowledge.

Yes, I know, knowledge is good, as is the search for new truths. We have been seeking knowledge since ancient times. Aristotle opened his *Metaphysics* with the simple statement: "All men by nature desire to know." We have, as a bedrock value in our society, long agreed on the value of open access to information, and recognize the problems that arise with attempts to restrict access to and development of knowledge. In recent times, we have come to revere scientific knowledge.

But despite the strong historical precedents, if open access to and unlimited development of knowledge henceforth puts us all in clear danger of extinction, then common sense demands that we reexamine even these basic, long-held beliefs.

It was Nietzsche who warned us, at the end of the nineteenth century, not only that God is dead but that "faith in science, which after all exists undeniably, cannot owe its origin to a calculus of utility; it must have originated *in spite of* the fact that the disutility and dangerousness of the 'will to truth,' of 'truth at any price' is proved to it constantly." It is this further danger that we now fully face — the consequences of our truth-seeking. The truth that science seeks can certainly be considered a dangerous substitute for God if it is likely to lead to our extinction.

If we could agree, as a species, what we wanted, where we were headed, and why, then we would make our future much less dangerous — then we might understand what we can and should relinquish. Otherwise, we can easily imagine an arms race developing over GNR technologies, as it did with the NBC technologies in the twentieth century. This is perhaps the greatest risk, for once such a race begins, it's very hard to end it. This time — unlike during the Manhattan Project — we aren't in a war, facing an implacable enemy that is threatening our civilization; we are driven, instead, by our habits, our desires, our economic system, and our competitive need to know.

I believe that we all wish our course could be determined by our collective values, ethics, and morals. If we had gained more collective wisdom over the past few thousand years, then a dialogue to this end would be more practical, and the incredible powers we are about to unleash would not be nearly so troubling.

One would think we might be driven to such a dialogue by our instinct for self-preservation. Individuals clearly have this desire, yet as a species our behavior seems to be not in our favor. In dealing with the nuclear threat, we often spoke dishonestly to ourselves and to each other, thereby greatly increasing the risks. Whether this was politically motivated, or because we chose not to think ahead, or because when faced with such grave threats we acted irrationally out of fear, I do not know, but it does not bode well.

The new Pandora's boxes of genetics, nanotechnology, and robotics are almost open, yet we seem hardly to have noticed. Ideas can't be put back in a box; unlike uranium or plutonium, they don't need to be mined and refined, and they can be freely copied. Once they are out, they are out. Churchill remarked, in a famous left-handed compliment, that the American people and their leaders "invariably do the right thing, after they have examined every other alternative." In this case, however, we must act more presciently, as to do the right thing only at last may be to lose the chance to do it at all.

As Thoreau said, "We do not ride on the railroad; it rides upon us"; and this is what we must fight, in our time. The question is, indeed, Which is to be master? Will we survive our technologies?

We are being propelled into this new century with no plan, no control, no brakes. Have we already gone too far down the path to alter course? I don't believe so, but we aren't trying yet, and the last chance to assert control — the fail-safe point — is rapidly approaching. We have our first pet robots, as well as commercially available genetic engineering techniques, and our nanoscale techniques are advancing rapidly. While the development of these technologies proceeds through a number of steps, it isn't necessarily the case — as happened in the Manhattan Project and the Trinity test — that the last step in proving a technology is large and hard. The breakthrough to wild self-replication in robotics, genetic engineering, or nanotechnology could come suddenly, reprising the surprise we felt when we learned of the cloning of a mammal.

And yet I believe we do have a strong and solid basis for hope. Our attempts to deal with weapons of mass destruction in the last century provide a shining example of relinquishment for us to consider: the unilateral U.S. abandonment, without preconditions, of the development of biological weapons. This relinquishment stemmed from the realization that while it would take an enormous effort to create these terrible weapons, they could from then on easily be duplicated and fall into the hands of rogue nations or terrorist groups. The clear conclusion was that we would create additional threats to ourselves by pursuing these weapons, and that we would be more secure if we did not pursue them. We have embodied our relinquishment of biological and chemical weapons in the 1972 Biological Weapons Convention (BWC) and the 1993 Chemical Weapons Convention (CWC).[12]

As for the continuing sizable threat from nuclear weapons, which we have lived with now for more than fifty years, the U.S. Senate's recent rejection of the Comprehensive Test Ban Treaty makes it clear relinquishing nuclear weapons will not be politically easy. But we have a unique opportunity, with the end of the Cold War, to avert a multipolar arms race. Building on the BWC and CWC relinquishments, successful abolition of nuclear weapons could help us build toward a habit of relinquishing dangerous technologies. (Actually, by getting rid of all but 100 nuclear weapons worldwide — roughly the total destructive power of World War II and a considerably easier task — we could eliminate this extinction threat.[13])

Verifying relinquishment will be a difficult problem, but not an unsolvable one. We are fortunate to have already done a lot of relevant work in the context of the BWC and other treaties. Our major task will be to apply this to technologies that are naturally much more commercial than military. The substantial need here is for transparency, as difficulty of verification is directly proportional to the difficulty of distinguishing relinquished from legitimate activities.

I frankly believe that the situation in 1945 was simpler than the one we now face: The nuclear technologies were reasonably separable into commercial and military uses, and monitoring was aided by the nature of atomic tests and the ease with which radioactivity could be measured. Research on military applications could be performed at national laboratories such as Los Alamos, with the results kept secret as long as possible.

The GNR technologies do not divide clearly into commercial and military uses; given their potential in the market, it's hard to imagine pursuing them only in national laboratories. With their widespread commercial pursuit, enforcing relinquishment will require a verification regime similar to that for biological weapons, but on an unprecedented scale. This, inevitably, will raise tensions between our individual privacy and desire for proprietary information, and the need for verification to protect us all. We will undoubtedly encounter strong resistance to this loss of privacy and freedom of action.

Verifying the relinquishment of certain GNR technologies will have to occur in cyberspace as well as at physical facilities. The critical issue will be to make the necessary transparency acceptable in a world of proprietary information, presumably by providing new forms of protection for intellectual property.

Verifying compliance will also require that scientists and engineers adopt a strong code of ethical conduct, resembling the Hippocratic oath, and that they have the courage to whistleblow as necessary, even at high personal cost. This would answer the call — fifty years after Hiroshima — by the Nobel laureate Hans Bethe, one of the most senior of the surviving members of the Manhattan Project, that all scientists "cease and desist from work creating, developing, improving, and manufacturing nuclear weapons and other weapons of potential mass destruction."[14] In the twenty-first century, this requires vigilance and personal responsibility by those who would work on both NBC and GNR technologies to avoid implementing weapons of mass destruction and knowledge-enabled mass destruction.

Thoreau also said that we will be "rich in proportion to the number of things which we can afford to let alone." We each seek to be happy, but it would seem worthwhile to question whether we need to take such a high risk of total destruction to gain yet more knowledge and yet more things; common sense says that there is a limit to our material needs — and that certain knowledge is too dangerous and is best forgone.

Neither should we pursue near immortality without considering the costs, without considering the commensurate increase in the risk of extinction. Immortality, while perhaps the original, is certainly not the only possible utopian dream.

I recently had the good fortune to meet the distinguished author and scholar Jacques Attali, whose book *Lignes d'horizons* (*Millennium,* in the English translation) helped inspire the Java and Jini approach to the coming age of pervasive computing, as previously described in this magazine. In his new book *Fraternités,* Attali describes how our dreams of utopia have changed over time:

"At the dawn of societies, men saw their passage on Earth as nothing more than a labyrinth of pain, at the end of which stood a door leading, via their death, to the company of gods and to *Eternity.* With the Hebrews and then the Greeks, some men dared free themselves from theological demands and dream of an ideal City where *Liberty* would flourish. Others, noting the evolution of the market society, understood that the liberty of some would entail the alienation of others, and they sought *Equality.*"

Jacques helped me understand how these three different utopian goals exist in tension in our society today. He goes on to describe a fourth utopia, *Fraternity,* whose foundation is altruism. Fraternity alone associates individual happiness with the happiness of others, affording the promise of self-sustainment.

This crystallized for me my problem with Kurzweil's dream. A technological approach to Eternity — near immortality through robotics — may not be the most desirable utopia, and its pursuit brings clear dangers. Maybe we should rethink our utopian choices.

Where can we look for a new ethical basis to set our course? I have found the ideas in the book *Ethics for the New Millennium,* by the Dalai Lama, to be very helpful. As is perhaps well known but little heeded, the Dalai Lama argues that the most important thing is for us to conduct our lives with love and compassion for others, and that our societies need to develop a stronger notion of universal responsibility and of our interdependency; he proposes a standard of positive ethical conduct for individuals and societies that seems consonant with Attali's Fraternity utopia.

The Dalai Lama further argues that we must understand what it is that makes people happy, and acknowledge the strong evidence that neither material progress nor the pursuit of the power of knowledge is the key — that there are limits to what science and the scientific pursuit alone can do.

Our Western notion of happiness seems to come from the Greeks, who defined it as "the exercise of vital powers along lines of excellence in a life affording them scope."[15]

Clearly, we need to find meaningful challenges and sufficient scope in our lives if we are to be happy in whatever is to come. But I believe we must find alternative outlets for our creative forces, beyond the culture of perpetual economic growth; this growth has largely been a blessing for several hundred years, but it has not brought us unalloyed happiness, and we must now choose between the pursuit of unrestricted and undirected growth through science and technology and the clear accompanying dangers.

It is now more than a year since my first encounter with Ray Kurzweil and John Searle. I see around me cause for hope in the voices for caution and relinquishment and in those people I have discovered who are as concerned as I am about our current predicament. I feel, too, a deepened sense of personal responsibility — not for the work I have already done, but for the work that I might yet do, at the confluence of the sciences.

But many other people who know about the dangers still seem strangely silent. When pressed, they trot out the "this is nothing new" riposte — as if awareness of what could happen is response enough. They tell me, There are universities filled with bioethicists who study this stuff all day long. They say, All this has been written about before, and by experts. They complain, Your worries and your arguments are already old hat.

I don't know where these people hide their fear. As an architect of complex systems I enter this arena as a generalist. But should this diminish my concerns? I am aware of how much has been written about, talked about, and lectured about so authoritatively. But does this mean it has reached people? Does this mean we can discount the dangers before us?

Knowing is not a rationale for not acting. Can we doubt that knowledge has become a weapon we wield against ourselves?

The experiences of the atomic scientists clearly show the need to take personal responsibility, the danger that things will move too fast, and the way in which a process can take on a life of its own. We can, as they did, create insurmountable problems in almost no time flat. We must do more thinking up front if we are not to be similarly surprised and shocked by the consequences of our inventions.

My continuing professional work is on improving the reliability of software. Software is a tool, and as a toolbuilder I must struggle with the uses to which the tools I make are put. I have always believed that making software more reliable, given its many uses, will make the world a safer and better place; if I were to come to believe the opposite, then I would be morally obligated to stop this work. I can now imagine such a day may come.

This all leaves me not angry but at least a bit melancholic. Henceforth, for me, progress will be somewhat bittersweet.

Do you remember the beautiful penultimate scene in *Manhattan* where Woody Allen is lying on his couch and talking into a tape recorder? He is writing a short story about people who are creating unnecessary, neurotic problems for themselves, because it keeps them from dealing with more unsolvable, terrifying problems about the universe.

He leads himself to the question, "Why is life worth living?" and to consider what makes it worthwhile for him: Groucho Marx, Willie Mays, the second movement of the *Jupiter Symphony*, Louis Armstrong's recording of "Potato Head Blues," Swedish movies, Flaubert's *Sentimental Education*, Marlon Brando, Frank Sinatra, the apples and pears by Cézanne, the crabs at Sam Wo's, and, finally, the showstopper: his love Tracy's face.

Each of us has our precious things, and as we care for them we locate the essence of our humanity. In the end, it is because of our great capacity for caring that I remain optimistic we will confront the dangerous issues now before us.

My immediate hope is to participate in a much larger discussion of the issues raised here, with people from many different backgrounds, in settings not predisposed to fear or favor technology for its own sake.

As a start, I have twice raised many of these issues at events sponsored by the Aspen Institute and have separately proposed that the American Academy of Arts and Sciences take them up as an extension of its work with the Pugwash Conferences. (These have been held since 1957 to discuss arms control, especially of nuclear weapons, and to formulate workable policies.)

It's unfortunate that the Pugwash meetings started only well after the nuclear genie was out of the bottle — roughly fifteen years too late. We are also getting a belated start on seriously addressing the issues around twenty-first-century technologies — the prevention of knowledge-enabled mass destruction — and further delay seems unacceptable.

So I'm still searching; there are many more things to learn. Whether we are to succeed or fail, to survive or fall victim to these technologies, is not yet decided. I'm up late again — it's almost six A.M. I'm trying to imagine some better answers, to break the spell and free them from the stone.

Notes

1. The passage Kurzweil quotes is from Kaczynski's Unabomber Manifesto, which was published jointly, under duress, by *The New York Times* and *The Washington Post* to attempt to bring his campaign of terror to an end. I agree with David Gelernter, who said about their decision: "It was a tough call for the newspapers. To say yes would be giving in to terrorism, and for all they knew he was lying anyway. On the other hand, to say yes might stop the killing. There was also a chance that someone would read the tract and get a hunch about the author; and that is exactly what happened. The suspect's brother read it, and it rang a bell. "I would have told them not to publish. I'm glad they didn't ask me. I guess." (*Drawing Life: Surviving the Unabomber.* Free Press, 1997: 120.)
2. Garrett, Laurie. *The Coming Plague: Newly Emerging Diseases in a World Out of Balance.* Penguin, 1994: 47–52, 414, 419, 452.
3. Isaac Asimov described what became the most famous view of ethical rules for robot behavior in his book *I, Robot* in 1950, in his Three Laws of Robotics:

    1. A robot may not injure a human being, or, through inaction, allow a human being to come to harm.
    2. A robot must obey the orders given it by human beings, except where such orders would conflict with the First Law.
    3. A robot must protect its own existence, as long as such protection does not conflict with the First or Second Law.

4. Michelangelo wrote a sonnet that begins:

   Non ha l' ottimo artista alcun concetto
   Ch' un marmo solo in sè non circonscriva
   Col suo soverchio; e solo a quello arriva
   La man che ubbidisce all' intelleto.

   Stone translates this as:

   The best of artists hath no thought to show
   which the rough stone in its superfluous shell
   doth not include; to break the marble spell
   is all the hand that serves the brain can do.

   Stone describes the process: "He was not working from his drawings or clay models; they had all been put away. He was carving from the images in his mind. His eyes and hands knew where every line, curve, mass must emerge, and at what depth in the heart of the stone to create the low relief." (*The Agony and the Ecstasy.* Doubleday, 1961: 6, 144.)
5. First Foresight Conference on Nanotechnology in October 1989, a talk titled "The Future of Computation." Published in Crandall, B. C. and James Lewis, editors, *Nanotechnology: Research and Perspectives.* MIT Press, 1992: 269.

   See also www.foresight.org/Conferences/MNT01/Nano1 .html.
6. In his 1963 novel *Cat's Cradle,* Kurt Vonnegut imagined a gray-goo-like accident where a form of ice called ice-nine, which becomes solid at a much higher temperature, freezes the oceans.
7. Kauffman, Stuart. "Self-replication: Even Peptides Do It." *Nature,* 382, August 8, 1996: 496. See www.santafe.edu/sfi/People/kauffman/sak-peptides.html.
8. Else, Jon. *The Day After Trinity: J. Robert Oppenheimer and The Atomic Bomb* (available at www.pyramiddirect.com).
9. This estimate is in Leslie's book *The End of the World: The Science and Ethics of Human Extinction,* where he notes that the probability of extinction is substantially higher if we accept Brandon Carter's Doomsday Argument, which is, briefly, that "we ought to have some reluctance to believe that we are very exceptionally early, for instance in the earliest 0.001 percent, among all humans who will ever have lived. This would be some reason for thinking that humankind will not survive for many more centuries, let alone colonize the galaxy. Carter's doomsday argument doesn't generate any risk estimates just by itself. It is an argument for revising the estimates which we generate when we consider various possible dangers." (Routledge, 1996: 1, 3, 145.)

10. Clarke, Arthur C. "Presidents, Experts, and Asteroids." *Science*, June 5, 1998. Reprinted as "Science and Society" in *Greetings, Carbon-Based Bipeds! Collected Essays, 1934–1998*. St. Martin's Press, 1999: 526.
11. And, as David Forrest suggests in his paper "Regulating Nanotechnology Development," available at www.foresight.org/NanoRev/Forrest1989.html, "If we used strict liability as an alternative to regulation it would be impossible for any developer to internalize the cost of the risk (destruction of the biosphere), so theoretically the activity of developing nanotechnology should never be undertaken." Forrest's analysis leaves us with only government regulation to protect us — not a comforting thought.
12. Meselson, Matthew. "The Problem of Biological Weapons." Presentation to the 1,818th Stated Meeting of the American Academy of Arts and Sciences, January 13, 1999 (minerva.amacad.org/archive/bulletin4.htm).
13. Doty, Paul. "The Forgotten Menace: Nuclear Weapons Stockpiles Still Represent the Biggest Threat to Civilization." *Nature*, 402, December 9, 1999: 583.
14. See also Hans Bethe's 1997 letter to President Clinton, at www.fas.org/bethecr.htm.
15. Hamilton, Edith. *The Greek Way*. W. W. Norton & Co., 1942: 35.

TED KERASOTE

A Killing at Dawn

FROM *Audubon*

IN THE VERY FIRST LIGHT, no more than a pale wash of sky, the four wolves move over the sage-covered hillside above Soda Butte Creek in the far northeastern corner of Yellowstone National Park. Here, the Absaroka Mountains of Wyoming and the Beartooth range of Montana meet, the Stillwater and the Clark Fork head, and if you walk away from this single road, you walk for several days before crossing another. With spotting scopes, binoculars, and radio-telemetry gear, we are watching the Druid pack, the wolves that have made this corner of Yellowstone their home.

"Freed of gravity" more accurately describes how the wolves move than the "trotting" or "loping" gaits they intermingle effortlessly as they hunt up valley. They seem like vapors, like shadows, tracing and mottling the contours of the ground — the occasional draw and swale and stand of aspen not impeding their travel in the least as they search for elk calves.

It is spring in Yellowstone, and the newborn calves — spotted, scentless, and weighing about thirty pounds — dot the moist green meadows, though we see none of them this morning as they lie in the sage, immobile as mushrooms, quiet as a heartbeat, banking on their immobility and the flood of newborn elk to keep them safe. As survival strategies go, it's a sound one. With thousands of calves dropped during the same few weeks in May and June, the chance of any individual calf's being caught by the wolves is small. Nevertheless, a few of them will be caught. In the conifers, a mile above this hillside, the Druid pack has six new pups to feed, and is motivated.

The four wolves hunt in a zigzag pattern: a black alpha male; two subdominant animals, one gray, one dark like her father; and, bringing up the rear, a more grizzled yearling male, as large as his older siblings but inexperienced. The increasing light now gives them definition: tails drooped, noses to the ground, backs arched, like a foursome of animated croquet hoops, raising their heads to pant and grin, their tongues flashes of wild rose in the dawn. They seem happy at their work, which isn't surprising. Evolution wouldn't have selected those who found hunting drear business.

Grizzly bears — there must be two dozen in this valley — are also hunting elk calves this morning. We can see them through our spotting scopes, in the treeline meadows, bird-dogging through the sage and trying to flush a potential meal. The females with cubs, however, lack the unhurried, *sans souci* attitude of the Druid pack. They charge here and there with frantic desperation, anxious to make a kill and return to their cubs, whom they've stashed a ways off, for a male bear could come along and wipe out their litters.

It isn't that the wolves are less concerned parents than grizzlies; rather it is that they live in a rich social network, and other pack members remain at the den to guard the pups. This morning the alpha female has stayed behind.

Watching the two species raise their offspring, I find it hard not to think of the differences in the human world between the difficulties single parents often face rearing children alone and the easier time some couples have, especially those who can call upon an extended family for baby-sitting. Such an analogy may sound anthropomorphic to the "zooseparatists" among us, but those not so narrowly inclined have long noticed that mammals share many traits and behaviors. Charles Darwin considered that nonhuman animals had "moral qualities," the more important of which were "love, and the distinct emotion of sympathy."

Before the recovery plans spawned by the Endangered Species Act reintroduced wolves to Yellowstone and also increased the region's grizzly bear population, and before the two species, in turn, forced elk to respond to the historic levels of predation that had shaped them and that had been missing from the ecosystem for sixty years, it was difficult to see the crossover between these three species and ourselves. It was also hard to see how slim the line can be that separates us from other members of the animal kingdom.

Now, without leaving the park's roads, one can step across that line on many a spring morning.

With a bleating screech, so childlike in its terror that it raises a shiver on my spine, an elk calf bolts skyward from the sage, a chestnut-colored shape — legs, neck, head — tangled in the jaws of the lead wolf. In an instant, the struggling calf is swamped by paws and lashing tails. As the dust rises, the yearling wolf lags behind. The calf, without any personality beyond its single haunting scream, comes apart like a rag doll — its haunch stolen away by the gray wolf, its foreleg scuttled off by her darker sibling, its abdominal and thoracic cavities emptied of viscera so quickly by the alpha male that it seems as if the carcass has been vacuumed clean.

Then, from the aspens fifty yards off, an elk charges, sleek as a sorrel mare, neck hair bristling, eyes bugging, snorting — and too late. At least to save what was her hidden calf. But not to drive off the wolves. No fools, they retreat. The largest of them weighs about 130 pounds. The elk weighs 500 and has hooves as sharp and effective as splitting mauls. She stands over the bloody pile of flesh and bone, sweeping her head threateningly in the direction of the wolves, who, like commandos confronted by a more heavily armed foe, have retired a hundred yards off and lie in the sage, waiting and watching.

Suddenly, the mother elk stiffens, sniffs at the carcass, and trots down valley, head back and mouth agape. She goes no more than a dozen yards before wheeling and returning, only to bypass her calf and dash up valley, crazed and disoriented. The wolves rush back and reclaim their meat, dragging the carcass behind a fallen aspen tree, where they gnaw at it. The elk stands at a distance and stares at them malevolently. Then she races at the wolves, scattering them in three directions.

We watch as the sun rises and the yearling wolf, a teenager that hasn't participated in the kill or dismemberment, saunters down through the sage, approaching to within forty yards of us. He lies in the long grass up to his ruff, his radio collar hidden and his back turned to the drama taking place above him. He gazes directly at us as I focus my spotting scope. His head appears in astonishing clarity: greenish-yellow eyes illuminated warmly by the flood of sunlight; the commas of his nostrils incised deeply into his moist black snout; his whiskers long and pearly; a fly settling on one of his furry ears. Glancing up at his pack and then back to us, the wolf minces his eyes in an expression I have seen on the face of my own dog countless times: Feel my pain. He knows he will get none of this meat. The wolf turns his head away and once again stares up the hill to where the mother elk stands protectively over her dead calf, and the other members of his pack wait beneath her.

The morning fattens. Cackling garrulously, a V of Canada geese flies downriver, the silence that settles behind them deeper for how they broke it. Two sandhill cranes trill from a gravel bar, and a coyote appears in the grass behind us. It jigs along the bank, stops momentarily, and studies the waiting wolves and the disputed carcass. Long reddish ears pricked forward, eyes piercingly alert, pointy muzzle tickling information out of the breeze, it calculates the odds of stealing some scraps. It continues on its way. After all, perhaps only fifteen pounds of meat, hide, and bone remain of the elk calf, and the wolves won't hesitate to kill a coyote.

The yearling wolf rises from the grass, extends his forelegs in a lavish bow, and trots off in the direction of his den. Two of the waiting wolves follow, leaving the black alpha male lying in the sage and watching the mother elk with the patience of stone. After a few minutes she makes her way up the hillside, searching back and forth, every one of her dazed movements seeming to signify that a mistake has occurred and that she might yet find her calf alive and well.

When she strays twenty-five yards from the carcass, the alpha wolf flashes into motion — black stone to sprinting blur in less than a second. He snatches up the calf in his jaws and tears away, the elk wheeling and barreling down on him, but he has judged the distance, the angles, and his ability to escape perfectly. He stays four yards ahead of her, expending no more energy than necessary. Then, as if realizing the futility of her pursuit, she skids to a halt and he lopes away with her mangled offspring in his jaws.

He runs several hundred yards before coming to rest in the grass. Glancing over his shoulder, he begins to nip at the calf with tender little bites. The mother elk stares at him, then retraces her route up the hillside, sniffing here and there before coming to the spot where blood stains the bunchgrass. She stops directly over the site of the kill, looks back to the wolf, and begins to grunt mournfully, her sides contracting and her muzzle elongating into the shape of a trumpet. A few moments later her bellow of loss and frustration floats down the hillside to us. Again and again she calls.

The black wolf glances one more time at the grieving elk before standing and getting the calf comfortably set in his jaws. He trots in a straight line toward the forest and his den. He has made his meat, and six new pups are waiting to be fed.

We watch the elk watching the wolf disappear into the trees, and she continues to cry out, turning this way and that, sending her dirge in every direction as the morning heat rises. I would like to see how long she remains there, but we have to head down valley to find other wolves.

On our return at sunset, fifteen and a half hours after her calf was taken by the Druid pack, we see the mother elk standing on the very spot it was killed, a monument to fidelity in a natural world that barely blinks at such recyclings of protein. She looks weary and beaten, her head at half-staff. She also appears immovable in her resolve to guard the site, or to stand witness to what has occurred, or to continue to hope for her calf's reappearance. Who can know what is in her mind, except perhaps another mother elk? Perhaps a wolf, determined to bring meat back to his pups, might know.

BARBARA KINGSOLVER
AND STEVEN HOPP

Seeing Scarlet

FROM *Audubon*

PICTURE A SCARLET MACAW: a fierce, full meter of royal red feathers head to tail, a soldier's rainbow-colored epaulets, a skeptic's eye staring from a naked white face, a beak that takes no prisoners.

Now examine the background of your mental image: Probably it's a zoo or a pet shop, with not a trace of the truth of this bird's natural life. How does it perch or forage or speak among its kind without the demeaning mannerisms of captivity? How does it look in flight against a blue sky? Few birds that inhabit the cultural imagination of Americans — north and south — are so familiar and yet so poorly known.

As biologists who have increasingly turned our work toward the preservation of biodiversity, we are both interested in and wary of animals as symbols. If we could name the thing that kept pushing us through Costa Rica in our rented jeep, on roads unfit for tourism or good sense, it would have been, maybe, macaw expiation. Some sort of penance for a lifetime of seeing this magnificent animal robbed of its grace. We wanted to know this bird on its own terms.

As we climbed into the Talamanca Highlands on a serpentine, pitted highway, the forest veiled the view ahead but promised something, always, around the next bend. We were two days and sixty miles south of San José, in a land where birds live up to the extravagances of their names: purple-throated mountain gems, long-tailed silky flycatchers, scintillant hummingbirds. At dawn we'd witnessed the red-green fireworks of a resplendent quetzal as he burst from his nest cavity trailing his tail-feather streamers. But no trace of scarlet yet, save for the scarlet-thighed dacnis (yes, just his thighs — not his feet or legs). We had navigated through an eerie morning mist in an elfin cloudforest and at noon found ourselves among apple orchards on such steep slopes they seemed flung there instead of planted. All of it was wondrous, but we'd not yet seen a footprint of the beast we'd come here tracking.

Then a bend in the road revealed a tiny adobe school, its bare-dirt yard buzzing with small, busy children. The *Escuela del Sol Feliz* took us by surprise in such a remote place — but in Costa Rica, where children matter more than an army, every tiny hamlet has at least a one-room school. This one had turned its charges outdoors for the day in their white and navy uniforms, and the schoolyard seemed to wave with neat nautical flags. The children carried tins of paint and stood on crates and boxes, painting a mural on the school's stucco face: humpbacked but mostly four-legged cows, round green trees festooned with round red apples, fantastic jungles dangling with monkeys and sloths. In the center, oversize and unmistakable, was a scarlet macaw.

This portrait of the children's environment was a study in homeland, combining important features of real and imaginary landscapes, and while their macaw had more dignity than Long John Silver's, he was still a fantasy. These children had picked apples and driven the family cow across the road, and some may have seen monkeys. But it's unlikely that a single one of them had ever laid eyes on a macaw.

Once, it was everywhere, in the lowlands at least, on both the Pacific and Atlantic coasts of this country. But in recent decades *Ara macao* has been pushed into a handful of isolated refuges as distant as legend from the School of the Happy Sun. Its celebrity in the school's mural cheered us, because it seemed a kind of testimonial to its importance in Costa Rica's iconography, and to the scattered, growing efforts to teach children here to take their natural heritage to heart. We'd come in search of both things: the scarlet macaw, and some manifestation of hope for its persistence in the wild.

Our destination was Corcovado National Park, on the Osa Peninsula, where roughly 1,000 scarlet macaws constitute the most viable Central American population of this globally endangered bird.

The Osa is one of two large Costa Rican peninsulas extending into the Pacific; both are biologically rich, with huge protected areas and sparse human settlements. Corcovado, about one tenth the size of Long Island, is the richest preserve in a country known for biodiversity. Its bird count is nearly 400 species; its 140 mammals include all six species of cats and all four species of monkeys found in Central America. It has nearly twice as many tree species as the United States and Canada combined. The park was established by executive decree in 1975, but its boundaries weren't finalized for nearly a decade, until after its hundreds of unofficial residents could be relocated. Hardest to find were the gold prospectors — who had a talent for vanishing into the forest — and the remnant feral livestock, though the latter disappeared gradually with the help of jaguars.

For us, Corcovado would be the end of a road that was growing less navigable by the minute. Our overnight destination was Bosque del Cabo, a private nature lodge at the southern tip of the peninsula, and our guidebook promised we'd cross seven small rivers to get there. We hadn't realized we'd do it without the benefit of bridges. At the bank of the first river, we plunged in with our jeep, fingers crossed, cheered on by a farmer in rubber boots leading his mule across ahead of us.

"This will be worth it," Steven insisted when we reached the slightly more treacherous-looking second river. No bridge in sight, no evidence that one had ever existed, though a sign advised: PUENTE EN MAL ESTADO — bridge in a bad state. Yes, indeed. The code of Costa Rican signs is a language of magnificent understatement; earlier in the trip we were informed by a sign posted on a trail up a live volcano: "Esteemed hiker, a person could sometimes be killed here by flying rocks."

Over the river safe and sound, with Golfo Dulce a steady blue horizon on our left, we rattled on southward through small *fincas* under the gaze of zebu cattle, with their worldly wattles and huge, downcast ears. Between farms the road was shaded by unmanicured woodlots, oil-palm groves, and the startling monoculture of orchard-row forests planted for pulp. Seedeaters and grassquits lined the top wires of the fences like intermittent commas in a run-on sentence.

At dusk, with seven rivers behind us, we pulled into the mile-long driveway of Bosque del Cabo under the darkening canopy of rain forest. The road tunneled between steep, muddy shoulders, but we could smell the ocean. Our headlight beam caught a crab in the road, dead center. We slid to a stop and scrambled out for a closer look at this palm-size thing. A kid with a box of Crayolas couldn't have done better: bright purple shell, red-orange legs, marigold-colored spots at the base of the eyestalks. We dubbed it the "resplendent scarlet-thighed crab" and nudged it out of the road. But we immediately encountered more, and suddenly we were seriously outnumbered. Barbara surrendered all dignity and walked ahead of the jeep at a crouch, waving her arms, but as crab-herd she was fighting a losing battle against a mile-long swarm. These land crabs migrate mysteriously in huge throngs between ocean and forest, and on this moonlit night they caught us in a pulsing sea of red that refused to part. They danced across the slick double track of their flattened fellows, left by drivers ahead of us. We've rarely traveled a longer, slower, *crunchier* mile than that one.

We slept that night in a thatched *palapa,* lulled by the deep heartbeat of the Pacific surf against the cliff below us. At first light we woke to the booming exchanges of howler monkeys roaring out their ritual "Here I am!" to position their groups for a morning of foraging. We sat on our little porch watching a coatimundi poking his long snout into the pineapple patch. A group of chestnut-mandibled toucans sallied into a palm, bouncing among the fronds. No macaws, though we were in their range now. We walked out to meet this astonishing place, prepared for anything — except the troop of spider monkeys that hurled sticks from the boughs and leaped down at us using their prehensile tails in a Yankee-go-home bungee-jumping display. Retreating toward our lodge, we heard a parrotish squawk in the treetops that we recognized from pet shops. Were they macaws?

"Sí, guacamayos," we were assured by a gardener we found shaking his head over the raided pineapple patch. Yes, he'd been seeing macaws lately, he said, usually in pairs, *"Practicando a casarse"* — practicing to be married. This was April, the beginning of the nesting season. Following courtship rituals, the macaw pairs would settle into tree cavities, always more than 100 feet above the ground, and lay their two-egg clutches. The young stay with the adults for as

long as two years; no more nesting occurs until after they have dispersed. This combination of specialized habitat and slow reproduction makes macaws especially vulnerable to an assembly of threats. The ravages of aerial pesticide spraying have lately diminished, as banana companies leave the country or switch to oil-palm production, but deforestation remains a phenomenal peril. Of the macaws' original Costa Rican habitat, only 20 percent still stands, all of it now protected. In addition to the Osa population, some 330 birds survive in the Carara Biological Reserve, to the north; scarlet macaws are also found in scattered pockets from southern Mexico into Amazonian Brazil.

Dire habitat loss has become the norm for tropical species, but macaws and parrots are further doomed by their own charm. The price of beauty is high for a young scarlet macaw captured by a poacher, who can sell it into the pet trade for as much as $400. (The fine for being caught is about $325.) Since 1990, when the nearby town of Golfito was allowed to begin collecting lower taxes on goods that come through its port, employment from the import trade has grown and poaching has declined noticeably. Farther north, however, in the economically undeveloped Carara region, poaching is ubiquitous. Many local conservationists feel the best hope is to develop alternative sources of income while educating children about the long-term tradeoffs of poaching, which could extinguish a national emblem before they're old enough to become adept at climbing 100-foot trees. During our trip, we spoke with several educators whose programs aim specifically at developing a family conscience about stealing baby parrots and macaws from their nest holes — revising a culture in which these birds have traditionally been harvested with no more moral qualms than a hungry coatimundi brings to a pineapple patch.

El que quiera azul celeste, que le cueste, the Costa Ricans say — If you want the blue sky, the price is high. The mix of hope and fatalism in this *dicho* speaks perfectly of the macaw's fierce love of freedom and touching vulnerability. We stood on the cliff near our *palapa* above the ocean, scanning, hoping for a glimpse of scarlet that wasn't there. Today we would complete our pilgrimage to Corcovado, where we would see them flying against the blue sky or we would not. On a trip like this, you revise your hopes: If we saw even one free bird, we decided, that would be enough. We prepared to push on the final ten miles to the road's end at Carate, gateway to the Corcovado forest, home to the country's last great breeding population of scarlet macaws.

Carate, although it appears on the map, is not a town. It's a building. Mayor Morales's ramshackle *pulpería* serves the southwestern quadrant of the peninsula as the singular hub of commerce: He'll arrange delivery-truck passage back out to Puerto Jiménez, buy gold you've mined, watch your vehicle while you hike, or simply offer a theoretical restroom among the trees out back. Indoors, dangling by wires from the ceiling, is a dazzling collection of bottles, driftwood, birds' nests, car parts — the very definition of *flotsam and jetsam,* if you can tell what floated in and what was jettisoned. Above the main counter dangles the crown jewel of the collection: a mammalian vertebra of a size generally seen only in museums. Under this Damocles bone we purchased a soda and plotted our strategy for finding macaws. Outside, on benches under a tree, we sat among the *pulpería* regulars, who explained to us that there are no roads into the park, no hiking trails, no wooden signboard maps declaring that you are here. Corcovado is not the user-friendly kind of national park we're accustomed to. How do you get in? You walk, and watch out for snakes. It's a thick jungle; where's the best walking? The beach.

While we chatted, a pet spider monkey sidled up to Barbara. Steven focused the camera. A barefoot girl nearby watched intently.

"Is he friendly?" Barbara asked in Spanish.

The girl grinned broadly. "*Muerde.*" He bites.

Steven snapped the photo we now call "interspecific primate grimace."

The steep gray beach offered rugged access to the park. The surf pounded hard on our left as we hiked, and to our right the wall of jungle rose steeply up a rocky slope. A series of streams poured down the rocks from the jungle into the Pacific. At the forest's edge the towering trees were branchless trunks for their lowest 100 feet or so. From this sparse, lofty canopy we began to hear macaws — not the loud, familiar croak but a low conversational grumbling among small foraging groups. We jockeyed for a view, catching glimpses of monkeylike movement as they clambered around, pulling fruits from clusters at the tips of branches. Macaws are seed predators, cracking the hearts of fruit seeds or nuts. High above the ground is where you'll see them, only and always, if you don't want bars for backdrop. Both the scarlet and the other Costa Rican macaw, the great green, rely on large tracts of mature trees for foraging, roosting, and nesting.

It's hard to believe something so large and red could hide so well in foliage, backlit by the tropical sky, but they did. We squinted, wondering if this was it — the view we'd been waiting for. Suddenly a pair launched like rockets into the air. With powerful, rapid wingbeats and tail feathers splayed like fingers, they swooped into a neighboring tree and disappeared uncannily against the branches. We waited. Soon another pair, then groups of three and five, began trading places from tree to tree. Their white masks and scarlet shoulders flashed in the sun. A grand game of Musical Trees seemed to be in progress as we walked up the beach counting the birds that dived between trees.

All afternoon we walked crook-necked and open-mouthed in awe. If these creatures are doomed, they don't act that way. *El que quiera azul celeste, que le cueste,* but who could buy or possess such avian magnificence against the blue sky? We stopped counting at fifty. We'd have settled for just one — we thought that's what we came for — but we stayed through the change of tide and nearly till sunset because of the way they perched and foraged and spoke among themselves, without a care for a human's expectation. What held us there was the show of pure, defiant survival: this audacious thing with feathers, this hope.

VERLYN KLINKENBORG

The Best Clock in the World

FROM *Discover*

IN A LAB off a humdrum hallway of a federal building on the western edge of Boulder, Colorado, an extraordinary instrument tosses a microscopic ball of cesium atoms up and down, up and down. I can see it happening. On a small monitor, a sudden, globular condensation of light flashes again and again as six lasers shape a cloud of cesium atoms into a tight sphere, then loft it upward into a cavity where it pauses and falls. The video feed is coming from somewhere deep within a cylindrical metal pillar in the middle of the lab. The pillar, as tall as I am, stands on a steel bench a little larger than a Ping-Pong table and perforated with holes. Mounted on the bench is a bewildering maze of lenses and mirrors designed to refine laser beams and shunt them into the pillar at various angles. The bench is shrouded in transparent plastic sheets attached to a frame of pressure-treated wooden studs — a crude dust shield. Another layer of plastic clings to the ceiling, which leaks from time to time. The contraption — bench, pillar, lenses, mirrors, plastic, studs, and all — looks like a model railroad, just waiting for track to be laid down and a train to start running through this brittle wilderness of glass and metal. A $650,000 model railroad, that is.

This device is actually a clock. In fact, NIST-F1, a cesium fountain clock housed at the National Institute of Standards and Technology, is the most precise clock in the world, a distinction it shares with a similar device at the Laboratoire Primaire du Temps et des Frequences in Paris. Despite its homemade look, F1 is accurate to 0.0000000000000015 of a second, or, as the scientists here put it, 1.5 parts in 10 to the 15th power. In other words, if it were to run for 20 million years, it would neither lose nor gain a second. Everyone who visits the Boulder lab instinctively looks at his or her wristwatch, hoping, somehow, to set it to this remarkable standard. I look at my watch, then at the fountain clock, then back at my watch. It's a naive, almost comical gesture. Ultimately, the output of the fountain clock has a bearing on the time my watch tells. But the difference between the two clocks is a story in itself, a story about the science of keeping time.

For centuries, Earth was our timekeeper. The sun rose and set, and the day was parsed into hours, minutes, and seconds based on Earth's rotation, which is why astronomical observatories, like the one in Greenwich, England, kept official time. Until the twentieth century, pendulum clocks were calibrated against the rotation of Earth by taking astronomical measurements. But as clocks grew more and more precise, they exposed the idiosyncrasies of our planet. It wobbles, it oscillates, it undergoes slight shifts in shape, all of which affect its rotation. As a standard of accuracy, the pendulum gave way in the 1940s to electrically induced vibrations in quartz crystals, which in turn gave way in the 1950s to measurements of atomic activity. And, in effect, Earth gave way to the atom as a gauge of time. Instead of using a definition of the second based on Earth's rotation, scientists began to search for one based on frequencies generated by certain atoms — particularly cesium — as they changed from one atomic state to another. Atomic frequencies, unlike the frequency of a pendulum's swing, have the virtue of being the same anywhere in the universe. In 1967, the international definition of the second shifted to an atomic standard. The Bureau International des Poids et Mesures (BIPM) near Paris now defines a second as "the duration of 9,192,631,770 periods of the radiation corresponding to the transition between the two hyperfine levels of the ground state of the cesium 133 atom."

It's one thing to define the second — to postulate its length — and something entirely different to build an instrument that can actually measure it according to that definition. Imagine the difference between defining a car and actually building one. Almost any enclosed vehicle that propels itself and carries passengers can be called a car. But the definition of a second is absolute, and what the scientists in the Time and Frequency Division at Boulder have done is build devices that come closer and closer to measuring the precise length of a real second, as defined by international convention. In the twentieth century, we have progressed from a 1921 pendulum clock, accurate to .3 second per year, to the fountain clock — 7 million times more accurate.

Gaining or losing a second here or there might not seem so important in our daily lives. But as the scientists at Boulder have chased greater and greater accuracy, industry has come to depend more and more on the increasingly precise measures of time that the Time and Frequency Division makes possible. Perhaps the best example is GPS, the Global Positioning System. Even an ordinary handheld GPS receiver determines its position by measuring the time it takes to receive signals from GPS satellites overhead. To put it simply, the precision of the coordinates your GPS receiver gives you — whether you're canoeing the Boundary Waters or surveying a new interchange on I-70 — depends directly on the precision of the time signal the system uses.

One of the classic scenes in any old war movie is the moment when the platoon leader gathers his men around him and says, "Let's synchronize our watches." We take it for granted that to carry out any coordinated action — military or civil — we need to agree on the time. We've spent the last century building increasingly complicated networks of machines and systems — satellites, Internet nodes, electrical grids, landline and cell-based phone communications. For those networks to carry out coordinated actions, they, too, need to agree on the time as they communicate with one another. In order to share information, computers in a network — whether it's an office network or the Internet as a whole — need to know when to talk, when to respond, and at what rate to do so. And, in a sense, the amount of information a network can distribute is directly related to how fast that information can be transmitted and how accurately time can be synchronized across a network. Time is no longer merely the natural fourth dimension of the universe around us. It's the beat that meters the electronic motions of money and information. Its precise measurement and distribution is a subject of detailed international agreements and of a treaty, called the Convention of the Meter, that has been in force since 1875.

Before the advent of railroads in the mid-1800s, it scarcely mattered if every city and every country kept time on its own, because the rate of communication among them was usually measured in days, weeks, and months. But these days a commonly distributed, accurately calibrated pulse of time, precise down to the level of billionths of a second, is what makes synchronization — indeed, communication itself — possible. When the public clock in a town in Renaissance Europe failed, it mainly affected the repairman who was called in to fix it. If our public clock failed, the Internet would dissolve into an array of freestanding, no-longer-networked computers. Trade would abruptly cease on Wall Street, and money and shares would come to rest wherever they were. Air traffic would stagger to a halt. Scientific labs of every description would find themselves deprived of their most fundamental measure. Even a temporary desynchronizing in the twenty-first century would leave turmoil in its wake. We would see in a sudden, displeasing instant how vitally our lives are shaped by agreement on the smallest measurable part of a second.

So, here I am, standing in front of the new fountain clock in Boulder, Colorado. This is, as they say, a primary standard, as far as you can presently go toward the measurement of time as a quantity. I look at my watch, which was set to the time on my computer at home, which was set by an automated link via the Internet to the time service at the Time and Frequency Division — just one of 25 million time requests that Boulder receives daily. The question is: "Am I looking at the source of that time?"

The answer is no. A cesium fountain clock isn't a tool for measuring the flow of time as it ticks past, month by month, year by year. It's a tool for measuring the length of the second, which is a very different thing. The fountain clock at Boulder usually runs for only a few days at a time. Each run is a test of sorts to see if the timekeepers can nudge a little more accuracy from the instrument, the way Formula One mechanics try to nudge one more mile per hour from their race cars. Time, as generated by this instrument, is an experimental result, an effort to match, in reality, an unwavering physical description of a second. But the second that the cesium fountain clock measures is any second, not a particular second from the flow of time. And even if this clock could tell you which second was passing — could tell you the time, in other words — it would still be an ambiguous result.

Don Sullivan, who is chief of the Time and Frequency Division, explains: "If you look at a single clock, you don't know whether it's right or wrong. And if you have two clocks and they disagree, you don't know which one's right. Things are much better with three clocks, because then you have majority voting." In other words, the cesium fountain can tell you the length of the second to a stupefying degree of precision, but it cannot, by itself, tell you the time in Boulder any more accurately than my ordinary wristwatch. In fact, it can't tell you the time at all.

What matters in international timekeeping isn't a single ultraprecise clock but a global network of clocks. The time that is propagated by international agreement — Coordinated Universal Time — depends as much on the accurate comparison of time measurements as it does on the accurate measurement of time itself. To keep accurate time, as Sullivan describes it, you need at least three clocks. If they're all in one room, it's easy to compare them. But if they're in different rooms, or in different buildings, or in different countries, how do you compare them, then? How, for that matter, do you formulate an "international" time?

The Time and Frequency Division maintains an entire array of clocks, called a time scale, not just the atomic primary standard. The fountain clock provides the measure of the second against which the rate of inherently less accurate but more stable clocks is judged. In another room, Sullivan shows me five hydrogen masers — a different kind of atomic clock — or rather he shows me the large commercial incubators, the size of refrigerators, that permanently house them. "These are our best reference clocks," he says. "They're very, very stable over short to intermediate times, intermediate being out to many months." Like precision — the ability to measure the second in microscopic intervals — stability is a critical factor in good timekeeping. Several different effects cause hydrogen masers to drift. The drift is minuscule but it's also predictable, which means that when the scientists at Boulder factor the drift out, they're left with a rock-steady record of time, the standard of steadiness against which other clocks in the time scale are measured.

For accuracy over longer periods of time, the National Institute of Standards and Technology relies on a roomful of commercial cesium clocks, the kind that anyone can buy for, say, $40,000 to $60,000. In fact, the time that comes from Boulder — what Sullivan calls "our best estimate of the international time" — is an average of the output of those cesium clocks plus that of the hydrogen masers plus that of the cesium fountain clock plus that of its predecessor, with which it is still running in tandem. "When we average all the clocks together, we weight them according to their long-term and short-term stability," Sullivan says. "The masers dominate the performance of the time scale out to months, but the cesium standards then start to become more important longer term. So it's a very intelligent average. The fountain clock then calibrates this periodically." What the time scale generates is a sophisticated average — an equation of time, so to speak — to which each kind of clock contributes its strength but not its weakness. It's as though one kind of clock added the seconds, another the minutes, and still another the hours.

Sullivan leads the way into one more room with rows of racks lined with stationary batteries — car batteries, essentially, but with the life span of a seventeen-year cicada. It is here that the peculiar isolation of the Time and Frequency Division becomes apparent. It would be unwise for a source of precise timing for national and international telecommunications to depend completely on the power grid. The time service Sullivan and his colleagues provide is so vital that frequency — even electrical frequency — is actually generated at Boulder. All of the clocks in Boulder's time scale are powered by this congregation of batteries, which is backed up by two generators, a reservoir of fuel, and technicians wearing pagers. "You wouldn't even trust a switchover to batteries," says Sullivan. "You just run the clocks on batteries and keep charging the batteries all the time. You don't want 99.999 percent reliability. You want 100 percent."

When it's noon sharp in Boulder, the Coordinated Universal Time is 7:00 P.M., or 6:00 P.M. during daylight-saving time. But international time isn't just Boulder time shifted six or seven hours ahead. It's an average generated in Paris from some 220 clocks around the world, including nearly a dozen primary standards, which are located at Boulder, the Paris Observatory, and labs in Germany, Japan, and Canada. Just as the National Institute of Standards and Technology time, roughly speaking, is an average of the clocks in its time scale, international time is an average of time scales around the world. In effect, it's the result of clock-comparison on a global scale, using a global tool: GPS and its Russian counterpart. Not surprisingly, the technique for international time comparison and distribution, called GPS common-view time transfer, was invented at Boulder.

Throughout the day and night, for thirteen minutes at a time, Boulder and all the other metrology labs in the world lock on to one GPS receiver after another and compare their clocks. The data is compressed and then brought together in Paris. And this is where a search for the source of time begins to grow very strange. "Weeks and weeks after the fact," Sullivan says, "Paris sends us a notice, and they say, 'Here is where each of your clocks is relative to the average.'" Boulder then tries to steer its own time-scale average to the average generated in Paris. *Steer* is an odd word to hear in the world of timekeeping. When I ask Sullivan what he means by it, he says, "One of the axioms in modern timekeeping is that you never touch a clock once it's running. So you never go in and say, 'Well, that clock's running a little fast.'" He reaches into midair to fiddle with an imaginary dial. The atomic standards and hydrogen masers and cesium clocks stay locked away.

Instead, what the Boulder lab steers is a computerized output based on the results from Paris. Sullivan calls it a paper clock, but, in fact, one of the clocks at Boulder is adjusted to track the international average. The time computer-users get when they log on to the Time and Frequency Division time server is its most accurate estimate, which is derived from a weeks-old international average based, in part, on its own, still-older initial average. Meanwhile, the Time and Frequency Division keeps a taped record of the individual output of all the clocks in its time scale so that, as statistical methods improve, scientists there can go back through time and analyze the accuracy and drift of those clocks. Gauging the past performance of the clocks helps the scientists predict their future performance. Somewhere in the laboratories at 325 Broadway, Boulder, Colorado, is a chronicle of every second that has passed since the National Institute of Standards and Technology established the time scale in 1960.

I had thought that in the presence of the cesium fountain clock I might have come to the farthest threshold of time. But other labs at Boulder are working on the possibility of newer, more precise clocks. Another threshold in timekeeping is already in sight, although it may be years before it's crossed. In the fountain clock the cesium atoms are observed while in motion, which makes their frequencies hard to determine. "We're keen on something called stored-ion frequency standards," says Sullivan. "The fountain clock — all of our clocks — are limited by the time that we get to observe the transition and the speed of the atoms. In stored-ion clocks, rather than slow the atoms down and look at them for a second, we just stop them altogether and trap them and hold them. We can look at the atoms indefinitely. In a sense it's an ultimate kind of technology." As the clocks at Boulder grow still more precise, they will reach a point where the very noise they generate while running becomes a limiting factor. The effect of gravitation itself will weigh in. In a sense, it already does. "Our clock runs at a different rate here in Boulder than it would at sea level, and we have to correct for that when we're doing our evaluation," says Sullivan. "Curiously, if we continue with the present development rate of clocks, ten or twenty years from now we will run into a position where knowing our location in the gravitational field will be one of the key difficulties." The clocks are already so precise that the effect of relativity can be detected within a gravitational field.

But no matter how radically the standard of timekeeping changes, no matter how minutely scientists manage to divide the second or how perfectly they distribute time itself, the questions St. Augustine posed some 1,600 years ago — when the minute, not to mention the second, had not yet been invented — will still prevail. "While we are measuring it," Augustine asked, "where is it coming from, what is it passing through, and where is it going?" These questions echo in nearly everyone's experience, whether they live by the clock in this split-second world or blithely measure their days by the sun and the moon. Mankind has made a science of measuring time and distributing it, but not of using it wisely. And when it comes to Augustine's questions, science still does not have any answers.

JON R. LUOMA

The Wild World's Scotland Yard

FROM *Audubon*

ONE SPRING DAY, Bonnie Yates plucked a tiny inch-long hair from a luxurious and elaborately embroidered shawl. Carefully, she placed the hair onto a glass slide, then moved the slide onto one stage of her dual-stage microscope, a favorite tool of scientists who investigate crimes. Next to it she placed a slide bearing another tiny hair, a reference sample she keeps under lock and key in her lab. Peering at the hairs, side by side, through her microscope, she could see that the small, fine strand from the shawl was a perfect match for the reference sample. Both hairs were crinkled, brittle-looking, "zigzagged," as Yates says, "like a French fry." Each hair's inner structure is shaped like a cobblestone path. As Yates knew better than anyone in the world, the hairs proved that someone had committed a crime.

Yates's reference sample was collected from a stuffed chiru, a Tibetan antelope. The hair's shape and appearance are unique to that mammal, meaning that the hair from the shawl could have come only from a chiru, a critically endangered species that's protected by international treaty. While there were as many as a million chirus at the start of the twentieth century, the population has since dropped to about 70,000.

In 1998 wildlife agents in Hong Kong had seized the shawl and dozens of others from a dealer, convinced that they were so-called ring shawls made of shahtoosh, the luxurious and contraband wool of the chiru. So named because the entire delicate garment can be pulled through a wedding ring, shahtoosh shawls are typically woven in a northern province of India. Warm but featherweight and supremely soft, they have long been considered the ultimate in luxury and can retail for as much as $15,000.

Traffickers in shahtoosh garments sometimes suggest that the chirus conveniently shed their wool onto bushes. In fact, virtually all the fleece that enters the market is collected with bullets. To get enough shahtoosh for a single ring shawl, poachers mow down and skin as many as five chirus.

Yates is a forensic scientist for the Clark R. Bavin National Fish and Wildlife Forensics Laboratory, in Ashland, Oregon. Her workplace, a sprawling, nondescript concrete building, is the world's only full-service lab that focuses exclusively on crimes against wildlife, a service it provides, not only to agents of the U.S. Fish and Wildlife Service, which runs the lab, but also to state wildlife agencies and foreign governments.

When wildlife agents find an endangered animal that has been shot or poisoned, or if they believe they've caught a poacher, or when customs agents seize a bit of apparent contraband, such as elephant ivory, the evidence is sent here — the wild world's Scotland Yard.

With its small but highly specialized staff of thirty-two scientists and support workers and an annual budget of $1.7 million, the ten-year-old lab has become central to the prosecution of crimes against wildlife. There is certainly no shortage of crimes. In 1999 the lab handled more than 1,000 cases. Every weekday, a small parade of FedEx, UPS, and Postal Service delivery trucks appear at the lab's loading dock with more evidence. On one recent day, technician Elaine Plaisance unpacked a routine delivery: four large ice chests, each containing a dead, apparently poisoned, bald eagle; a black skull-and-crossbones poison symbol had been stamped on each cooler. From cardboard boxes lined with vinyl sheeting, Plaisance retrieved eighty oil-covered ducks; another box contained a lump of alligator meat. In another she found a wad of tissue that might have been the gall bladder of a bear, an organ so prized in traditional Asian medicine that illegal traffickers can make thousands of dollars poaching bears. The lab has perfected a test that proves that U.S. traffickers often substitute the similar-looking gall bladders of farm pigs, which is not a crime.

Most of the mountain of contraband that has come to the lab for analysis over the years is now stored in a Colorado warehouse, although Ken Goddard, the lab's director, keeps a few of the unique samples on display in what he calls the "shop of horrors." First Goddard shows off a tacky caiman purse, complete with the endangered reptile's head and hind legs, and then the "$6,000 version" — sleek and classically styled, its latticed skin dyed a silky black. "I probably shouldn't mention which fancy New York store the agents found this in," he says.

Next Goddard holds up a plaque bearing the head of a green turtle, and a pair of high-top tennis shoes made with the skin of a Russell's viper, both of which are endangered species. Goddard points to a stool made from an elephant's foot and then picks up a pair of ordinary-looking shoes. "Elephant leather," he says. And that, he adds, is the essence of the problem that wildlife agents — and the lab — often have to deal with. While the source of the stool is all too evident, only an expert would know that the leather is actually elephant hide.

"If federal agents drag a whole elephant in here, you and I and everyone else would know it's an elephant," Goddard says. "But we almost never get easily recognizable parts, much less a whole animal. Most often we get pieces — bones, skins, feathers, scrimshaw, dried blood, leather. All the things that might tell you 'this is an elephant' aren't there."

That's precisely why, before the lab opened its doors, federal, state, and international wildlife agents often struggled in vain to prosecute crimes against wildlife, including a flood of traffic in endangered species that's estimated to be worth billions of dollars annually. Much of the time, gaining a conviction was hopeless, because unlike crimes against humans, crimes against wildlife often leave behind no easily identifiable victim, only those often-baffling bits and pieces of evidence.

"It's impossible to even begin to express how important this lab has become to us," says Kevin Adams, the assistant director for law enforcement at the Fish and Wildlife Service headquarters, in Washington, D.C. He remembers sometimes begging conventional crime labs to help analyze evidence. "But there are limits to what a guy from the state police whose expertise is in hit-and-run accidents can do. Now we have the world's recognized experts working in our own lab. When it's time to take cases to court, we're getting convictions."

Richard Dickinson, a Fish and Wildlife special agent in Minnesota, cites an extraordinary 1997 case in which a Minnesota man killed a gray wolf, a threatened species. The man insisted that the wolf had been charging his six-year-old son. But since wolves are virtually always afraid of humans, wildlife agents were skeptical.

"You really couldn't tell by a cursory examination of the bullet wound what had happened," Dickinson says. "So before the lab existed, the guy would have probably gotten away with it." But the lab's chief wildlife pathologist, Richard Stroud, and its ballistics team were able to show that, as Dickinson puts it, "debris from the bullet in the wound was flowing from the back of the neck to the front. That meant we could prove in court that the wolf was running away from the guy when he shot it."

In the past, suspected poachers and smugglers often escaped prosecution simply because no expert could prove conclusively that a sample of bone, skin, or fiber was, in fact, from a protected species. For instance, before Bonnie Yates documented the shahtoosh's microscopic signature in 1996, merely distinguishing that wool from other fine wools, such as cashmere, was nearly impossible.

Not anymore. Early in 1999 Yates flew to Hong Kong to serve as the key expert witness in the high-profile trial of the textile trader whose shawls she had examined the previous year. Hong Kong wildlife agents had seized 130 of the contraband shawls in the trader's luxury hotel room. On the strength of Yates's testimony, the trader was convicted and fined $40,000. This past July Yates's analysis scored another victory, in a federal court in New Jersey, when import-export dealers Linda Ho McAfee and Janet Mackay-Benton pleaded guilty to violating the Endangered Species Act by exporting almost 100 shahtoosh shawls for sale in a fashionable Paris boutique.

According to Fish and Wildlife special agent Tara Donn, who worked on the case, "The defense never even contested the forensic results. The forensic work from the lab really was central to the case." Above all, she says, open trade in shahtoosh has come to a virtual stop in the wake of the lab's discovery, although she suspects that a black market remains active.

Paradoxically, despite the obvious and pressing need for such a facility, the nation had a wildlife-crime-lab director for a decade before it had a lab. Goddard had spent twelve years as a forensic scientist in conventional crime labs in California before answering an ad to run a national wildlife-forensics lab in 1978. But Fish and Wildlife hired him before it had the funding to build a laboratory, much less hire a staff. The running joke through most of the 1980s was that the national laboratory was located "in Ken Goddard's briefcase."

Partly to vent his frustrations, Goddard began churning out crime novels, full of the kind of grisly violence and police procedure he had been exposed to during his days with the sheriff's department. His first novel, *Balefire,* was a bestseller in 1983. In more recent books, like *Double Blind,* the heroes and villains are wildlife agents, poachers, and radical environmentalists. In a sort of *X-Files* turn, Goddard's most recent book features a forensic scientist who receives a sample of tissue from a truly exotic creature; the sample turns out to contain, as he puts it, "DNA, certainly, but not DNA of this world."

After Congress appropriated the funds, the lab finally opened in 1989. Today high-tech gear fills the facility. Moving from wing to wing of the building recently, Goddard waved at gadgets used to replicate and analyze DNA; an enormous scanning electron microscope, based on a design from Scotland Yard, which occupies a small room of its own; supercooled "ultra-freezers" containing about 30,000 samples of known animal tissues that are used to make DNA comparisons; and a gleaming, dishwasher-size device called a gas-chromatograph that's designed to sniff out the chemical makeup of virtually any compound.

But for all the technological wonders, Goddard says he doubts that computers, or any technology, can ever be substituted for human expertise. Pepper Trail, a staff scientist and the lab's ornithologist, taps his head as he puts it this way: "A high degree of computing power from up here is brought to bear on these identifications. This is still the most powerful instrument we rely on here."

Goddard and I find Trail seated on a low stool at a large square table in the busy and cluttered section of the lab devoted to morphology, the branch of biology that focuses on the form and structure of organisms. Here scientists spend their days comparing evidence with the lab's enormous collection of reference samples: feather to feather, skeleton to skeleton, claw to claw, bone to bone.

Trail is examining a collection of wing feathers that arrived in a recent shipment with a request for identification. He has arranged them on a large lab table according to their curvature, and has already concluded that they come from a garden-variety male pheasant.

The morphology section centers around an open, capacious room, brightly lit but ghoulish. Here, on shelves everywhere, on tables, in trays and in drawers and boxes, is a fantastical collection of reference samples: carefully preserved and labeled animal body parts, including bones, skulls, whole skeletons, beaks, claws, and talons, as well as entire stuffed mammals and birds. Surrounding Trail's worktable is tray after tray bejeweled with stuffed and glassy-eyed songbirds. On a shelf are similar-looking skulls, one labeled DOG, the other WOLF. On one table sits the skull of a gorilla, and on another there's a partially dissected wallaby, exuding the pungent odor of a butcher shop.

"I hope body parts don't bother you," says Goddard, who explains that the wallaby died of natural causes in a California zoo and is now in the process of joining the collection of reference skeletons. The skinned carcass, he notes matter-of-factly, will be left for a few days in the company of the lab's resident colony of "very thorough" dermestid, or flesh-eating, beetles, which will produce a clean skeleton.

Like any modern crime lab, the wildlife version often uses DNA analysis to identify species, or even individual animals. The lab is able to identify the DNA's unique genetic "fingerprint" in a snippet of tissue. Its DNA work, for instance, helped lead to the November 1999 conviction in a Brooklyn federal court of three smugglers, including the former deputy chief of police in Warsaw, Poland, who attempted to sneak thousands of pounds of sturgeon eggs (caviar) past Fish and Wildlife inspectors at Kennedy Airport. (To protect several endangered species of sturgeon, legitimate caviar importers follow strict permit and reporting regulations.) The case marked the first time that caviar smugglers had been prosecuted since regulations went into effect in 1998; one of the smugglers was fined $25,000 and sentenced to twenty months in federal prison.

Still, DNA analysis has its limits. Unlocking the genetic code of a tissue sample is complex and expensive. And although human DNA is now well characterized, knowledge about animals' DNA is still evolving. Besides, when it comes to the practicalities of dealing with lawyers and juries, sometimes it's much easier to gain a conviction with lower-tech approaches.

"It's faster and cheaper and easier to explain to a jury," Trail says of the morphological approach. "With complex technology, like DNA, a defense attorney can try to cloud and confuse the issue. But it's a compelling thing when I can hold up one of our known, labeled reference bones, or a feather or a talon, and hold up a matching bone or feather or talon, and show people that they're the same."

To prove his point, Trail takes a golden eagle foot from a drawer. "This is an easy one," he says. "Nothing else looks like it. It's too big for a hawk and is the only eagle we've got with feathers all the way down."

With a Ph.D. in ornithology and an extensive background in field research, Trail was well qualified when he was hired by the lab in 1998. But, he says, it took months of intensely studying feathers and bones on the job to develop his deep knowledge of the minutiae of bird morphology — enough to sail through the rigorous hands-on tests the lab gives every few months. Trail, for instance, might be given a few claws, feathers, and bits of bone and asked to identify the species they come from. Now one or two feathers is all he needs to identify a bird.

Mammalogist Bonnie Yates is legendary for her ability to recognize species from fragments — often with something as scant as a small piece of a tooth. One day in the lab's DNA section, puzzled scientists were struggling to identify a mysterious lumpy bone. Yates wandered by. "Hey, look at that cool giraffe vertebra," she exclaimed, cracking the case on the spot.

The forensic lab has been making breakthrough discoveries since it opened, in large part because there are simply so many unknowns. "Wildlife forensics is a science still very much in its infancy," says Edgard Espinoza, the lab's chief of science. "In some ways we're where human forensic science was 100 years ago. But that's what's fun about working here: We have the opportunity to push a new science out to places it hasn't been before."

Indeed, soon after the lab opened, Espinoza and staff scientist Mary-Jacque Mann solved a problem that had long baffled customs and wildlife agents: determining if an ivory artifact is carved from elephant ivory, which is illegal to import, or from woolly mammoth or mastodon ivory, which is legal both to possess and to import. When trafficking in elephant ivory was banned in 1989, poachers began labeling shipments "mammoth." Espinoza and Mann discovered that because of internal structural differences, tiny lines in the ivory vary with absolute consistency between the species. Today any customs agent with a magnifying glass, a plastic protractor, and a copy of the lab's guidelines can tell the difference.

In recent years the lab's scientists have developed a simple technique for distinguishing the gall bladders of bears from those of pigs by chemically evaluating bile from the organs. And its DNA specialists have found a way to extract the genetic material from rhinoceros horns, even though the horns are essentially composed of keratin, which does not break down easily. The secret: The scientists use liquid nitrogen to supercool the keratin to about minus 60 degrees centigrade, at which point it does break down.

Ballistics experts at the lab are also working on new ways to use computers to digitize images of spent bullets to better analyze their patterns. And in the late 1990s Espinoza himself made a landmark discovery when he found that the chemical composition of the proteins that constitute hemoglobin in blood are unique to each species. In most cases, the lab needs only a small speck of blood to identify a species conclusively.

Espinoza is happy to demonstrate the technique, using samples of dried blood from a loon and an eagle, and then some of his own blood, each captured on bits of gauze. He dissolves the dry samples in solvent in separate test tubes, then quickly separates the solvent containing the blood from the gauze in a whirring centrifuge. One by one, he pops a mote of blood from each test tube into a device called a liquid chromatograph mass spectrometer. For each sample, the chromatograph spits out a chart — an image of peaks and valleys that amounts to a map of the chemicals in the blood. He need only compare each chart with a series of reference charts. The matches are perfect: loon, eagle, and human.

The lab never comments on ongoing cases or on how its work might be used in future prosecutions. But judging by its record, it seems certain that it will keep pushing out new boundaries to put poachers, smugglers, and other wildlife criminals out of business.

CYNTHIA MILLS

Breeding Discontent

FROM *The Sciences*

ANIMALS BREED, MULTIPLY, and die on the computer screen in Jonathan Ballou's office at the National Zoo in Washington, D.C. Ballou, a population geneticist, is running a program designed to simulate what happens when the population of an endangered species — in this case, the golden lion tamarin, a small Brazilian primate — becomes so low that only a small group of animals is left.

The computer screen is filled with squares and circles: the squares are male tamarins, the circles females. The shapes contain pairs of letters that represent specific genes. The letters are paired because animals always have two copies of a gene: one from their mother, and one from their father.

Ballou starts the computer program, and cybermating begins. The existing circles and squares are randomly matched to generate a new row of circles and squares: the first generation. Those new shapes then quickly beget their own offspring: the second generation. The result is a virtual family tree, in which each shape displays its own particular combination of inherited letter pairs. That's important, because genes come in flavors called alleles. A gene always codes for the same trait — resistance to disease, for instance — but its alleles allow for variety: some alleles may confer more resistance than others. When a pair of letters on the screen don't match — *AB, CD* — the gene in question was inherited as a different allele from each parent. When the letters in a pair are the same — *BB* or *DD,* for instance — the animal received identical alleles from each parent.

For a while, Ballou's virtual propagation program proceeds at full tilt, and all seems well. Gradually, though, a subtle change emerges. The letter *C* disappears, then *A*, and eventually, like the dismal, flat electroencephalogram of a person who has just died, the variety on the screen is reduced to a repetitive monotone. Many animal icons remain, but they are all the same: shape after shape becomes *BB* or *DD*. Soon each animal is genetically identical to the next. To a geneticist that spells annihilation: even if many individual animals remain alive, the extinction of the species looms.

Computer models enable biologists to reduce the future of a species to mathematical formulas — a series of tightly controlled steps. Will the species survive, or will it go the way of the dodo and the passenger pigeon? Ballou can vary such parameters as population size, genetic makeup, and the average number of offspring produced by a mating, then sit back and watch as the virtual animals evolve.

One of the inevitable pressures on any species is genetic drift, the random reassortment of genes from one generation to the next. In large populations genetic drift can lead to fluctuations in the frequency of certain alleles, though the law of large numbers makes it highly unlikely that the genetic variations represented by any particular alleles will disappear completely. But small populations are unequipped to resist the ravages of genetic drift. An allele that happens not to get passed on by one mating pair simply drops out of existence. So, randomly and inexorably, alleles in the gene pool of endangered populations disappear, the populations lose their genetic diversity, and soon the remaining animals become too inbred to survive. The wrenching predicament for conservation biologists is that endangered species reach the point of no return before their numbers fall to zero. A species can be living, yet genetically doomed.

The last great hope that animal biologists have cherished in the fight to protect some critically endangered species is captive breeding. The idea is to bring animals into zoos, coax their numbers higher, and then release them into their native habitats. The vision, in fact, goes even further than that. The world, most biologists acknowledge, is in the midst of a biological holocaust. Even as the human population has surpassed the 6 billion mark, domesticating wilderness along the way, the number of pandas can be counted in the hundreds, and Florida panthers in the dozens. The remaining animals are at the mercy of chance: a hurricane, a war, an epidemic, or the simple toss of the genetic dice could wipe them out forever. All a biologist can hope for is that, with time, the human population will stabilize and the political will to restore and permanently protect endangered habitats will eventually prevail.

Captive breeding gives impetus to that scenario because it enables people to *do something* about extinction. It is a hands-on process, and it presents a set of technological problems to be solved. Biologists can design software to plan breeding strategies. And if they can't get the right animals to mate, they can turn to the latest reproductive technology. Now that sheep have been cloned, why not Siberian tigers? Some biologists are even working to create frozen zoos, stocked with the embryos of exotic animals.

But captive breeding has been problematic. Animals born in captivity and then released into nature often have trouble finding food, fleeing predators, and selecting mates. As a result, some of the same biologists who initially promoted the technique are now voicing serious doubts about its viability. And some critics even question whether the entire endeavor might somehow be misguided. In most cases, after all, the result is a species that survives only in zoos. What is a wild animal when it can no longer live in nature? Technically, the animal has been saved, but hasn't something else been lost — something that cannot be calculated, cannot be captured by geometric shapes on a computer screen?

Fifteen years ago, faced with unprecedented rates of species extinction, conservation biologists proposed a global captive-breeding plan as a desperate last resort. In a much-cited article published in the journal *Zoo Biology* in 1986, the conservation biologist Michael E. Soule, now an emeritus professor at the University of California, Santa Cruz, and a group of others in the field proposed what they called the "millennium ark." The so-called ark would be made up of captive populations of 925 endangered animal species, from Caribbean flamingos to Siberian tigers — species that would be doomed to extinction without the absolute protection of cages and compound walls. The goal was to support the animals for the two centuries or more that would be needed to restore and protect their habitats.

But despite its nickname, the project was no Noah's ark, with pairs of animals tucked into cozy stalls. The conservation biologists knew they had to have far greater numbers: at least fifty animals of each species would be needed simply to protect against the vagaries of nature, and population geneticists have calculated that between 500 and 5,000 animals would be necessary to safeguard a species' genetic heritage.

Zoo administrators took to the idea immediately. For one thing, it seemed doable: zoos had already shown that they could coerce various fussy species to breed. Graduating to endangered animal would be a natural next step. The project was bound to generate some good publicity, which would be welcome: zoos had recently come under fire from animal-rights activists. In addition, captive breeding was a good draw: babies, even ugly ones such as fledgling condors, make for great press releases.

The American Zoo and Aquarium Association (AZA) in Silver Spring, Maryland, worked hard to build a zoo-led captive-breeding effort that was based on solid scientific research. Zoos developed species- and habitat-protection plans, and they began to keep careful studbooks for every endangered species under their control. Zoo biologists carefully chose which animals would be bred, and they acted with an almost autocratic zeal: if tiger X was a good genetic match for tiger Y at another zoo, tiger X would be moved, no matter how fond the local schoolchildren might be of him.

Usually, captive breeding involves animals that are already living in zoos. But on rare occasions, as with the California condor, biologists have captured all the remaining representatives of a species in order to breed them in captivity. Biologists resort to such extreme measures only when they are certain the animals have no chance of survival in the wild.

All the snazziest tricks of reproductive technology have been enlisted in the cause. Indeed, endangered species undergo the same high-tech procedures as wealthy, infertile human couples do. Artificial insemination is one technique: semen is collected from a male and then injected into the reproductive tract of a female. Getting the semen can be complicated, however. Horse and cattle breeders often obtain it by inducing their animals to mount a dummy, but males of other species won't necessarily comply. One alternative, which has been carried out successfully with black-footed ferrets, is to place probes in the anuses of the males and apply electric shocks that force them to ejaculate.

Another technique is in vitro fertilization (IVF), in which sperm and eggs are mixed in a test tube and the resultant embryos are implanted in a female's uterus. Sometimes the sperm or eggs are even culled after a rare animal's death and implanted in a surrogate of the same species. The high cost of IVF limits its use to only the most endangered species. But for those species, no workable technology seems too expensive, and even such delicate and sophisticated manipulations as the injection of a single sperm into the cytoplasm of an egg have been tried. Embryos and unfertilized eggs are sometimes frozen, and cloning has been proposed as a future option.

In spite of all those efforts, however, the results of captive breeding have been disappointing. An interim review published in 1994 by Chris D. Magin and his colleagues at the World Conservation Monitoring Centre in Cambridge, England, found that only about 15 percent of the space devoted to mammals in zoos was filled by endangered species. And the total number of those animals was low: only about half those species were represented by populations of more than 50 individuals, and less than 7 percent of the species were represented by more than 500 animals.

As for reintroduction projects, in the same 1994 review, a team led by the conservation biologist Benjamin B. Beck of the National Zoo concluded that if success is defined as a population of 500 individuals living independently in the wild, then success was elusive. Out of 145 projects (involving the release of more than 13 million animals, the vast majority of them fishes), only 16 projects met those criteria (including 2 reintroductions of American bison that took place at the beginning of the twentieth century). Given that record, which many biologists agree has not improved much since the 1994 review, it is easy to feel disillusioned. And consider the costs of these projects: more than $250,000 a year for Puerto Rican parrots, between $500,000 and $1.5 million a year for California condors. Can such low success rates justify continued efforts?

Granted, there have been some victories. By 1972 the golden lion tamarin had been reduced to between 100 and 200 animals living in the wild. Devra G. Kleiman, a conservation biologist at the National Zoo, began a breeding program for the tamarins, with the participation of several zoos that already held captive populations. Biologists began reintroducing captive-bred tamarins in 1984, mostly on private ranches northeast of Rio. A major public-education campaign put the tamarin on T-shirts and posters, and made it a kind of Brazilian national mascot. Today there are about 800 tamarins living wild in Brazil, of which some 300 are descendants of reintroduced animals. Other success stories are the Arabian oryx, which was reintroduced into the Middle East beginning in 1982 (though poaching has since depleted its numbers again); and the black-footed ferret and the California condor, both of which have been brought back to the western United States.

But a distinguished group of zoo and wildlife biologists led by Noel F. R. Snyder, then at Wildlife Preservation Trust International in Philadelphia, pointed out that those cases have been the exceptions. In a controversial 1997 article in the journal *Conservation Biology,* the biologists argued that though captive breeding has seized the public's imagination and brought glamour to zoos, the technique is inappropriate for many species and should take a back seat to habitat conservation.

In the past two decades, in response to soaring extinction rates, biologists have devised an entirely new subdiscipline, the science of small populations. As Ballou's work demonstrates, the loss of alleles from the gene pool of a species is like the loss of stocks from a mutual fund. The resultant portfolio looks increasingly shabby, as issues that perform well only under certain risky scenarios are sold off, and the portfolio as a whole becomes increasingly vulnerable to random fluctuations in the market.

Similarly, when genetic diversity is lost, the overall health of an animal population declines: its capacity to respond to new insults becomes attenuated. Furthermore, harmful genetic variations are expressed more frequently because the genetic options have been narrowed: a harmful gene that is usually recessive in a large population is more likely to be paired with its twin in a small population during reproduction. The result is fewer offspring (many die as embryos) and populations of individuals that are so similar to one another that they can all be wiped out by a single chance event.

To counter those forces, population geneticists such as Ballou reduce sex to a science. They run through all the permutations, hypothetical mating animal W with animal X and animal Y with animal Z. Then they try to ensure that endangered animals are interbred in such a way that a minimum of genetic variation is lost.

Of course, all that planning assumes that you can put a male and a female in a cage and let nature take its course. Sometimes the animals do go right to work: Ballou says golden lion tamarins mate practically on request. More often, however, the animals are inconveniently choosy. "In a lot of these captive breeding situations, where the mate choice is basically computerized, you can expect to be disappointed," says Scott R. Derrickson, the curator of ornithology for the Smithsonian Institution in Washington, D.C., which runs the National Zoo.

Not all species refuse to breed in captivity; some exhibit just the opposite problem. Guam rails are a case in point. A handful of rails and rail eggs were rescued from the island of Guam in 1983, before the invading brown tree snake could eat them all. Once the birds got to the National Zoo's Conservation and Research Center in Front Royal, Virginia, they bred almost like chickens. Unfortunately, unlike chickens, rails are so antisocial that they kill one another unless they are kept in separate cages. All that space for one species gets to be expensive, and zoos clamored for the birds to be reintroduced into the wild. Between twenty and thirty rails are now set free every year on Rota, an island near Guam that is free of the brown tree snake, and recently, rails have also been released in an area of Guam from which the snakes have been removed.

Another major problem for captive breeding is that key information is often lacking. For example, Micronesian kingfishers were rescued from Guam around the same time as the rails were, but the captive population has barely held on, hovering at around thirty birds. And because the birds were not studied prior to their extinction in the wild, there is no way to determine what is inhibiting them from breeding in captivity. To address the problem, the wildlife ecologist Susan M. Haig of Oregon State University in Corvallis and one of her graduate students, Dylan C. Kesler, are studying a related subspecies of kingfisher that inhabits Pohnpei, an island near Guam. There, mated pairs depend on adult offspring to help them raise their chicks. In addition, mating is preceded by a courtship display in which the males fly up high and then dive-bomb tree-borne termite colonies, which the birds then appropriate as nests. The Guam kingfishers might become better breeders, therefore, if zookeepers kept them together as families and gave them attractive targets to dive-bomb.

The first rule of any reintroduction project is obvious: whatever killed off the population in the first place must be controlled. Habitat must be protected, predators (including humans) contained, pollution cleaned up, and invaders tossed out. Yet as simple as that principle is, it has been neglected, even in the most widely publicized efforts. California condors were dying, in part, because they were ingesting lead bullets from the carcasses of deer during hunting season. So the condors were trapped and taken into "protective custody," as Noel Snyder likes to call it. As soon as the birds were doing well in captivity, biologists began releasing their offspring back into the wild. Yet the birds still die from lead poisoning (as do golden eagles and vultures), because, in the areas where they fly free, neither hunting nor lead ammunition has been banned.

Another simple rule of thumb: the easiest species to save are the ones with the fewest neurons. If, like butterflies and snakes, animals are "hard-wired" — meaning that they enter the world knowing everything they need to know — there is no need to train or acclimatize them: just breed them and release them into a suitable habitat.

Many animals, however, need to learn complex behaviors if they are to forage successfully and avoid ending up as something else's dinner. Without exposure to older, more experienced members of their own species, young captive-bred animals are at a major disadvantage. They are like children raised without culture or tradition. They don't know how to be wild.

People have tried teaching such skills to captive-bred animals, but the results have been pretty dismal. For one thing, animals don't learn as well or as quickly from people as they do from members of their own species. And there is a more subtle but still devastating problem: even carefully screened exposure to people can make animals tame. When feeding captive condors, workers wore condor-shaped hand puppets in order to avoid habituating the birds to human contact — but the tactic didn't work very well. Unlike wild condors, puppet-raised condors are unafraid of people. They beg food from campers and have even traipsed nonchalantly into suburban houses.

Even more catastrophic for captive-bred animals is their inability to avoid predators. When a group of animals is released into the wild, as many as 80 percent of them may be eaten by other animals. In some particularly depressing cases, all the animals have been killed. Another major problem posed by captive breeding is disease. Zoo officials don't like to talk about it much, but zoos are potential breeding grounds for disease. They collect and house animals native to habitats that are continents apart, and those animals carry pathogens that are only too willing to take up with new hosts. Cockatoos from Australia have passed deadly beak and feather disease to macaws and Amazon parrots. Avian tuberculosis is another killer that is often passed among captive birds of different species. Sometimes the animals may be contagious without exhibiting any obvious symptoms. Thus even if quarantines are imposed, there is always a risk that captive-bred animals might spark an epidemic in the wild when they are set free. Snyder now believes that disease is the most serious threat to reintroduction efforts.

For some species — including Guam rails, Micronesian kingfishers, and black-footed ferrets — captive breeding is probably the only hope against the abyss of extinction. But one has to raise the question: What are we saving if the animals have lost their essential wildness? "Philosophically, what do we want?" Jesse Grantham, director of the Mississippi state office of the National Audubon Society, asks of the condor project. "We can save the species, that is certain. After that, is it simply to have the blood and guts of a live bird? Or is it something greater than that?"

As Grantham points out, an excessive focus on captive breeding can make people forget — or even undermine — the true goal: preserving *wild* animals. "Do we compromise the integrity of a species while preserving it, or do we insist on saving it in situ, the species and its interactions with other species and the environment?" he asks.

Indeed, many zoos have begun to do exactly what some biologists suggested years ago: they are focusing less on breeding and more on saving habitat and protecting animals in the wild. Undoubtedly, some biologists will continue to design frozen zoos filled with the cryo-preserved embryos of otherwise lost species. But honing technology can accomplish only so much. Wild animals cannot be reduced to shapes on a computer screen, or objects to be gawked at in a zoo. The irony of endangered species is that once they have been brought permanently into our world, they are lost to us forever.

OLIVER MORTON

Ice Station Vostok

FROM *Wired*

TO IMAGINE Lake Vostok, you must first envision a great lake in a living landscape, a week's walk from end to end, too wide to see across from the highest hills on its flanks. Now simplify, subtract. Erase the surrounding woods and fields; hide the encircling hills. Remove the changing seasons and the replenishing rain. Still the winds that drive the waves; silence the beaches that soften the shores. Take out the fish; take out the weeds. Finally, shut out the sky itself; turn a surface of blue reflection to one of dark constraint. Leave only the waters, the minerals, the muddy depths. Then trap them, squeeze them, estrange them from everything that lives and dies elsewhere. From your creation emerges a simple world, a world that hungers for more. About 150 miles long and 31 miles across, Lake Vostok is elongated in shape, much like eastern Africa's Lake Malawi or Siberia's Lake Baikal. In area (5,400 square miles), it's nearly the size of Lake Ladoga, the greatest of Russia's western lakes. Its volume (2,900 cubic miles) is a good match for Lake Superior's. Its maximum depth (1,800 feet) is similar to Lake Tahoe's.

These proportions place Vostok comfortably among the world's fifteen largest lakes — but its setting is utterly singular. Lake Vostok sits beneath 2.5 miles of solid ice, close to the center of East Antarctica. If it ever had a direct link with the air above it, that connection ended some 30 million years ago. Climatologists at the ragged Russian base near Vostok's southern end — three hours by plane from the South Pole — have recorded the coldest temperature ever measured on Earth's surface: minus 128.6 degrees Fahrenheit, nearly cold enough to freeze carbon dioxide out of thin air. The only visible rocks within 600 miles are shooting stars fallen to Earth, rare black pebbles in the icescape.

Lake Vostok's forbidding remoteness makes it supremely hard to reach, and therefore unbearably attractive to scientists. Vostok's very isolation means it holds secrets no other place on the planet can. Its sediments contain a unique record of Antarctica's climate before the ice caps arrived — one that could revolutionize the science of the frozen continent. There could be prehistoric life in its waters, an indigenous ecosystem surviving with few resources — no sunlight, the tiniest of fresh-food inputs — and spurring adaptations never seen before. Were Lake Vostok open to the rest of the world, its faint records and fragile life forms would have been overwritten long ago. But being cut off may have preserved them from the changes that endlessly erode and recreate the world above.

Vostok's existence was unknown until thirty years ago, when radar and seismographs allowed scientists to piece together a map. The first hints of water under the ice were detected in the 1970s; much later, in the early 1990s, satellites and data from earlier seismic surveys revealed Lake Vostok's full extent. In 1995, a borehole was drilled from Russia's Vostok station, named for the ship sailed by nineteenth-century Russian explorer Fabian Gottlieb von Bellingshausen and built in that godforsaken spot, quite by chance, long before anyone suspected something important might be below. The borehole came within 400 feet of entering the lake, but the drillers stopped short of breaking through to the waters beneath.

In the near future, however, millions of years of isolation may come to an end: Researchers from America, Britain, Europe, New Zealand, and Russia have started lobbying their governments for a multimillion-dollar, long-term effort to fathom Vostok's depths. Funding, for the most part, would come from the National Science Foundation in the United States, the Natural Environment Research Council in the United Kingdom, and various other European organizations working together under the Scientific Committee on Antarctic Research (SCAR), the international body that oversees science in the southernmost continent under the terms of the Antarctic Treaty.

If the multinational teams of scientists get their way, the exploration of Lake Vostok — perhaps the most ambitious and complex scientific undertaking Antarctica has yet seen — could begin in less than five years. New bases will be built, some temporary, some permanent; new logistical infrastructures will be created to serve them; specially engineered drills will bore holes in the ice with blasts of hot water; and fleets of aircraft will transport thousands of gallons of fuel oil. (It takes a hellish amount of energy to get through 2.5 miles of ice.) Tele-operated and autonomous deep-diving robots will launch themselves from the boreholes into the great lake's waters, then sink through the blackness to the silent ooze below.

Glaciologists, sedimentologists, biologists, geochemists, and climatologists will try to understand how the lake waters circulate, where they come from, and where they go. They'll look for prehistoric life forms and search for nutrient sources. They'll hunt for clues to how long the lake has been there, and what came before it. Long-dark Vostok will be pried open for inspection — a process that, however carefully undertaken, runs the risk of changing the lake forever and destroying what has made it unique.

Why take that chance? Some believe Vostok should be left alone because exploration might permanently damage its pristine ecosystem. But proponents of drilling believe Vostok could provide new insights into the young Earth's spectacular ecological crises, during which the whole planet was frozen solid, its oceans reduced to the very brink of lifelessness. And it could illuminate the possibilities of life farther off — in a vast ocean on Europa, Jupiter's fourth-largest moon, 483 million miles from the sun. Along with Mars, Europa is the most likely prospect for evidence of life beyond Earth, and Vostok is the closest thing to a Europan environment our planet has to offer. Isolated from light, warmed only from below, starved of nutrients, the life forms of Vostok could teach scientists how life might persist in Europa's frigid climate, where temperatures average minus 250 degrees Fahrenheit. It would certainly show them how to look for it there: Exploring Vostok would be the nearest thing to a space mission without leaving the planet.

It's no surprise, then, that at NASA's Jet Propulsion Laboratory in Pasadena, California, the probes that will look for life in these icy depths millions of miles apart are being developed side by side. At Caltech and JPL, which designs and operates most of NASA's planetary spacecraft, a team of six engineers is adapting instruments, first developed to explore the hot depths of ocean-floor volcanic vents, for use in the isolated waters of Vostok or in Europa's alien oceans. The team is trying to develop a small metal probe — a cryobot (*cryo* means "ice" in Greek) — that can emit chemicals to create a sterile zone around itself, burrow through the ice using a high-pressure jet of hot water to clear its path, take samples, and transmit collected data back to the surface. It's also devising ways for the probe to see, retrofitting deep-water cameras to squeeze through a narrow borehole in the ice and still function in conditions of extreme pressure and cold.

But it's not just the technology that links Vostok and Europa. It's the ideas — pure, powerful ideas — that take scientists from the everyday world to somewhere far stranger, more abstract. In Lake Vostok's dark waters, they hope for visions from halfway across the solar system, and halfway back to Earth's beginnings.

"The lake — isolated for several million years — could contain microbes that have evolved in a manner unique to this environment," says Martin Siegert, a University of Bristol glaciologist who worked on the early satellite research that first exposed Vostok. "In effect, the lake can be thought of as a laboratory for investigating evolution."

Everything runs down. Hot baths get cold; fires turn to ash. The heat of Earth's core is slowly leaking out to space. In Antarctica, though, this heat loss is stymied by an insulating barrier: an ice sheet weighing billions of tons. The top of the ice is very cold indeed. The bottom, gently heated from below, is in some places warm enough to melt and collect as liquid water in pools or lakes.

In the 1970s, American, British, and Danish researchers used 60-MHz airborne radar to hunt for such hidden water pockets in Antarctica. Because radio waves can travel through ice, signals from the radar bounced off not the ice sheet's top, but its bottom. In some places, radio waves reflected evenly, suggesting that the ice was sitting on something much smoother than craggy bedrock. The obvious explanation was a layer of water. In 1974, the multinational team identified the longest stretch of subglacial water ever seen, near the Soviet Union's Vostok base. It was the first glimpse of the lake.

Few found the discovery very exciting. "Nobody talked about bodies of water under the ice — nobody thought about them," says Cynan Ellis-Evans, a microbiologist with the British Antarctic Survey (BAS). "Everyone had this mental picture of the water being just a few tens of meters thick." But in 1991, Jeff Ridley, a remote-sensing specialist with the Mullard Space Science Laboratory at University College London, directed a European satellite called *ERS-1* (the first long-duration civilian radar) to turn its high-frequency array toward the center of the Antarctic ice cap. It confirmed the 1974 discovery: a flat plain at 105 degrees east by 77 degrees south — a glacial dimple nearly 150 miles long, with Vostok station at the southern end.

When Siegert put this new satellite imagery together with the airborne-radar readouts and older Soviet seismic charts, the vast lake was revealed in all its glory. "It was a eureka moment," says Siegert. Although Siegert had cataloged evidence for some seventy subglacial lakes, Vostok was by far the largest and most contained. "The data couldn't be interpreted as anything except a deep body of water," he says. And, purely by chance, other scientists had already drilled most of the way down to it. They weren't interested in the lake — they didn't know it was there — but instead were extracting and studying ice cores from the glacier above.

The ice below Vostok station is a history book, with each core a vertical record of prevalent conditions at a particular point in time: the amount of dust in the atmosphere, the levels of trace gases, even the weights of the water molecules themselves, which provide hints to major climatic events. The ice cores scientists had been pulling up told the story of four 100,000-year cycles, a century of the world's climate in each yard.

The Russians started drilling into this record in the 1970s, the French joined them in the early '80s, and the Americans signed up in 1989. By the time Siegert and his colleagues could confirm the lake's existence, almost two miles of ice had been hauled out of the glacier above, split into several slices, sampled, and shipped off to American, French, and Russian labs around the world. The scientists realized that if this routine continued, and drilling went on unchecked, they'd pierce the last level of ice and tap into Lake Vostok within a matter of months.

When Ridley made his findings public in 1993 in the *Journal of*

Glaciology, he immediately roused scientific interest, moving SCAR to suggest that a workshop be held to decide if continued drilling made sense. In May 1995, two dozen researchers gathered to discuss this question at the Scott Polar Research Institute in Cambridge, England — the same institute where the original radar work was masterminded. "The U.S., France, and Russia — the three nations involved in the drilling — asked SCAR to come up with a recommendation regarding Vostok, and were willing to accept the outcome as being for the greater good," Ellis-Evans recalls. In other words, if anyone was going to drill into Vostok, the committee had to give the green light.

News of the meeting made its way to Ellis-Evans at BAS headquarters on the other side of town. As a biologist studying Antarctica, Ellis-Evans already knew that its surface lakes mattered, but like most of the world, he had no idea that the subsurface Lake Vostok even existed. "Frozen surface lakes of the Antarctic continent are an oasis," he says, "sitting in a Martian-like landscape, teeming with life." That life is mostly in the form of photosynthesizing microbial mats on the lake beds. Freed from predators by Antarctica's isolation, these mats leach nutrients from the "gin-clear" water, says Ellis-Evans, and create a stratified microbial ecology in the sediments below.

But there weren't going to be any photosynthesizing mats in Vostok's darkness 2.5 miles down. The water would be colder than in the shallow surface lakes, and the pressure immense. That didn't mean there could be no life. Assuming Vostok had once been a surface lake before ice covered the continent, organisms from that ecosystem might have found a way to survive. If that were the case, scientists would be looking at the ultimate biological time capsule — one that dates back 30 million years. The organisms' origins — and their adaptations to a weird world of cold, dark scarcity — made them the most tempting of targets for study.

"What the glaciologists wanted to do at the time was get as long a climate record as possible," says Ellis-Evans. "Another few hundred meters could have given them another glacial cycle. But that would mean drilling down to the lake. They wanted to know if the rest of the Antarctic science community had a problem with that. We told them that the more we heard about this, the more concerned we were about it. And we outlined the possibilities for life. As we had our say, we could see they were changing their views — they saw that they could damage the biology irreparably."

The potential problem Ellis-Evans describes is quite simple: The pressure at the bottom of the ice sheet would squeeze any borehole shut unless it was filled with a nonfreezing fluid — in practice, dirty kerosene. Putting a borehole like that through the ice and into the lake would create a miniature *Exxon Valdez,* releasing toxic fluids in what could well be one of the most delicate, perfectly preserved ecosystems on Earth.

At the close of the three-day meeting in Cambridge, SCAR agreed that Lake Vostok should be given special protection, and the American/French/Russian team was asked to halt drilling within 400 feet of the lake's surface. "A country could go against SCAR," says Ellis-Evans, "but most nations work within the essentially apolitical organization and accept its ground rules." So far, the agreement has been honored. During the meeting, it became clear that SCAR's interest in protecting Vostok and the life it might harbor was another reason to study the lake further by less polluting means.

Shortly after the Cambridge mandate, Antarctic Treaty countries signed a multinational agreement to further protect the vulnerable continent. "Any work in Antarctica now has a firm legal requirement to recognize and address environmental concerns," says Ellis-Evans. But even with the new legislation, the science of Lake Vostok continued to be a strange dance between the desire to protect and the desire to study. As researchers became more intrigued with the lake, their conflicting desires to either save or spoil it were slowly being pulled together.

When the size of Lake Vostok — and the possibility of ancient life within it — dawned on the wider world, the excitement began to spread. "I'd pretty much ignored it until two years ago," says Robin Bell, a geologist at Columbia University's Lamont-Doherty Earth Observatory who supports a Vostok drilling mission. "I couldn't see why it was important. And then I realized that it's a leaping-off point for a geological understanding of East Antarctica. As unique geologic structures are identified, the biological questions of survival in a hostile environment will follow — the geology is, in essence, the cradle in which life can form. Vostok," she adds, "could be linked to an understanding of life on Earth millions of years ago and to the potential for life beyond Earth."

In 1998, scientists arranged further meetings — first in St. Petersburg, Russia, then in Washington, D.C. — to discuss the lake, and whether it was worth going to at all. Last fall, members of the growing Lake Vostok community returned to Cambridge to try to pull together a definitive statement about Vostok's scientific value. Such a declaration is essential if SCAR is to set up a formal Vostok committee, itself necessary to convince the American, European, and Russian governments that it's worth spending millions of dollars to fund a multinational exploration of the lake.

The conference was held on a rainy Sunday morning at the end of September, with sixty established experts and young hopefuls crammed into a meeting room at Cambridge University's Lucy Cavendish College. Among those on hand was Peter Barrett, a tall, soft-spoken New Zealander who is trying to pinpoint the onset date of Antarctica's glaciation; Gordon de Quincey Robin, a short, round Brit who pulled together the radar surveys showing the first clues of Vostok; and Roland Psenner of the University of Innsbruck, one of the world's foremost authorities on Alpine lakes. Though Psenner has studied Europe's high lakes — from Spitsbergen to Spain's Sierra Nevada — he's never gotten close to Antarctica. Still, he was thrilled by the possibility that Vostok might contain living microbes from a bygone era.

"Since my focus is on microorganisms living in ice and icy water," Psenner said, "Lake Vostok is a target of high priority, and I am convinced that certain forms of life exist in the lake. But before even considering a study, we must do our homework and find a secure, clean sampling technique."

As Ellis-Evans, the organizer, scurried in and out of the conference room, torn between collating the science and arranging the coffee breaks, Robin Bell kicked things off with her now-polished Vostok 101 talk, which detailed the lake's geological history and its potential as a subject for study. Jean Robert Petit, a glaciologist with the Laboratory of Glaciology and Geophysics of the Environment in Grenoble and the French leader of the team that studied the climate core taken from below Vostok station, described the crystal structures found in the ice. "From a glaciological point of view," he said, "Vostok is a unique place to study a very long paleoclimatic record." Valery Lukin, a Vostok expert from the Arctic and Antarctic Research Institute in St. Petersburg, talked about the latest seismic work carried out from Russia's base, data showing that the northern end of the lake is a shallow, rocky swamp, while in the south, the waters are deep — and the sediments below them possibly deeper still.

It took just three hours to cover all the facts known about Vostok. After lunch — and over the next twenty-four hours — scientists ran through the seemingly interminable list of unanswered questions. For example, when a break-out group debating biological issues quizzed Ian Dalziel, associate director of the Institute for Geophysics at the University of Texas, about the nature of the hole that the lake waters are sitting in, he cheerfully told them that it could have been created by two continents colliding, a landmass tearing apart, a glacier, or even an asteroid — and that it could have been there for as long 35 million years.

Another point of contention is whether Vostok was a surface lake that froze during a past ice age or something that formed more recently from water trickling through the Antarctic ice cap. Apparently, there's no way to tell whether Vostok's bedrock cradle already had water sitting in it before ice first took hold of Antarctica 35 million years ago. "It's easy to speculate," said Dalziel, "because there isn't any data. Until we drill it, we won't know for sure."

As luck would have it, the 1995 borehole for climate studies was drilled into an ice sheet whose bottom layer is made up of frozen water from Lake Vostok itself. Two teams at the Cambridge meeting had looked at samples of this refrozen water. One was headed by John Priscu of Montana State University, who has also researched the biology in Antarctica's surface lakes — which are frozen over but still receive enough light to perform photosynthesis. The other was led by David Karl of the University of Hawaii, an expert on marine microbiology who has observed life around deep ocean vents and in the nutrient-poor wastes of the North Pacific.

The two men — who have known each other for years — had no idea they'd both been working on Vostok problems until a few weeks before they presented their results at the Cambridge meeting. Priscu's and Karl's discoveries, published in separate papers in the December 10, 1999, issue of *Science,* concluded that refrozen ice above the lake contained quite a lot of bacteria. Karl's team reported that some of these bacteria are still able to function despite their very slow metabolisms.

That there could be living bacteria in the ice above Lake Vostok isn't much of a surprise in itself. Vostok may seem extraordinarily isolated to humans, but bacteria experience it as just another place to turn up. And just turning up is what bacteria do best: They do it all over the planet. Admittedly, if the bacteria were migrating into the ice above Lake Vostok, they would have spent more time getting there than most anywhere else — sinking slowly through the ice cap, each new year's worth of snow forcing them that little bit deeper. It would be quite possible for bacteria to melt out of the ice over one part of Lake Vostok, drift through its otherwise lifeless waters for a bit, then get frozen back into the ice elsewhere, waiting for their next chance to turn up someplace interesting. If that were true, the bacteria in Lake Vostok would be no different from those trapped in the ice above it.

But it's also possible that the microbes Karl and Priscu retrieved from the ice are part of a viable ecosystem indigenous to the lake, consisting of organisms that have found ways to get on with life in cold, dark, nutrient-poor water for millions of years. If such an ecosystem exists, the likelihood is that it's sparse and staggeringly slow to replenish itself. "I would say a generation every hundred years might be feasible, if the lake's as low in nutrients as we think," says Jim Tiedje, a professor of microbiology at Michigan State University who studies the types of organisms that make their home in Antarctic ice.

The problem facing life in Vostok is that if microbes can't photosynthesize, they must rely on chemicals in the environment that have the potential to react with each other and release energy. Stripped to basics, this normally means an organism will need to take electrons from whatever it uses as food and give them to oxygen atoms. For this to be sustainable, the organism needs a reliable supply of oxygen to accept the electrons, which, happily enough, Earth's plants are able to ensure. Some bacteria use alternative electron acceptors — sulfate ions, nitrate ions, carbon dioxide — but these, in turn, are mostly made available by oxygen-dependent life elsewhere. If you look into the sediments under one of the microbial mats Ellis-Evans studies in Antarctica's surface lakes, you'll see this cascade of chemical dependency beautifully illustrated, with each layer using byproducts from the one above: The ash from one biochemical fire becomes fuel for the next.

The microbial communities Karl studies around deep ocean vents — populations undreamed of a few decades ago — live in total darkness, but much of their energy use depends on electron acceptors that originate in the sunlit ecosystems far above. Without these imports from the world of photosynthesis, the ecosystems at the deep vents would have to get by on locally produced chemical imbalances that offer only a tiny fraction of the energy available to an ecosystem blessed with sunlight.

Some oxygen gets to Vostok through the ice — but most of it is likely trapped in little cages of water molecules called clathrates, and thus is probably unavailable to microbes. So if there's a microbial ecology in Vostok's depths, it's probably made up of the hungriest, slowest-growing microbes on the planet. To microbial ecologists like Tiedje, such a discovery would be fascinating in its own right. To earth scientists, it would offer new clues to the controversial "snowball-Earth" theory, which describes a series of catastrophes more than half a billion years ago that represent the most severe environmental crises ever.

It started when all life was microbial, the sun was dimmer than it is now, and the continents were clustered near the equator. Warm, wet soil absorbed more carbon dioxide than Earth's burden of life was producing. The greenhouse effect weakened, the world cooled, and ice spread over the oceans from the poles. Since more ice cover means more sunlight reflected back into space, things cooled down quickly. After the ice caps reached about 30 degrees latitude, the reflection effect became unstoppable, enveloping the entire planet in freezing temperatures and sending ice from both poles toward the equator. Every ocean, sea, lake, and river froze. With no sources of water vapor, clouds vanished from the sky. In the darkened oceans, photosynthesis shut down and the seas began to die. The snowball-Earth theory asserts that for roughly 10 to 50 million years the planet was reduced to a state close to death — probably very close to the state today of Vostok and, possibly, Europa.

While its seas were dying, though, Earth's nonbiological activities continued more or less as usual. Volcanoes pumped out fresh carbon dioxide, and since the newly frozen land was unable to absorb it, the gases built up in the atmosphere. Eventually, at the equator, the temperature rose back above freezing. The first cracks appeared in the ice. Water vapor, a powerful greenhouse gas, returned to the skies. That warmed things up further. More ice melted, more sunlight was absorbed, more water evaporated, and the temperature shot through the roof. Models suggest that it took only 100 years for the ice to recede back to the poles — a century during which the global mean temperature is thought to have increased by 122 degrees. Life leaped forth from its refuges in hot springs and lakes. With huge amounts of carbon dioxide and sunlight available, photosynthesis in the oceans ran riot, producing oxygen at an incredible rate. The wild warming after our last snowball Earth may have been the crucible from which the complex life forms of the Cambrian explosion arose.

The snowball-Earth theory features prominently in the report that Ellis-Evans, Bell, and others are preparing to submit to SCAR on the need for further research into Vostok. The report covers all the areas outlined in Cambridge: the presence or absence of an exciting ecology, the geological setting, the effect of the water flowing into the lake from the ice sheet above, the circulation patterns of water that is very cold and under extremely high pressure, the record of climate and ice movements in the sediments, and the source and fate of the water itself.

"A system like Vostok's, isolated in the cold and dark, would give us insights to the likely environment on Earth when snowball conditions would have prevailed," says Ellis-Evans. "It would tell us about what survived and how it survived. Nowhere else on Earth is likely to do that."

Ricardo Roura is one of the handful of people who have spent a winter on Antarctica without any governmental assistance. In the early 1990s, Greenpeace set up a camp at the edge of the Ross Sea as part of its campaign for the ratification of the Madrid Protocol to the Antarctic Treaty, which imposed a fifty-year moratorium on mineral exploitation. Roura was part of that camp, and his stay there taught him to love the barren land. As a coordinator for the Antarctic and Southern Oceans Coalition (ASOC), a group of 240 nongovernmental organizations from fifty countries with interests in the region, Roura spends much of his time trying to protect the Antarctic from environmentally disruptive projects. When it comes to Vostok, he doesn't think the scientists — particularly the NASA scientists — are taking those protections seriously. The exploration of Lake Vostok is just the sort of thing he wants to stop.

"Lake Vostok is to Antarctica what Antarctica is to the rest of the planet: remote, pristine, and unique," Roura told the Cambridge meeting. "You scientists have decided a priori that something must be 'done' with Lake Vostok as soon as technically possible, while the most appropriate option to protect the ultimate scientific and environmental value of Vostok would appear to be to postpone drilling the lake for the indefinite future." A particular bugbear for Roura is the idea that NASA wants to use Vostok as a test bed for technologies in development for Europa. He calls their interest selfish and shortsighted.

NASA puts a more positive spin on the situation. "One of the things we want to emphasize is that it really works both ways," says Joan Horvath, an MIT-trained rocket scientist and former manager of JPL's Europa-Vostok Initiative. "Because people are developing interesting technology that didn't exist before for space, we'll be able to go into the lake in a way that's clean and sterile and won't disturb it. You might not have been able to do that if you weren't trying to do both projects."

Nevertheless, activists like Roura and James Barnes, ASOC's general counsel and cofounder, continue to push for Vostok's protection. "In Vostok, we hope we can make the case for being cautious — not moving quickly to drill, considering all the options, and helping people understand what's at stake," says Barnes, who's also an attorney for Friends of the Earth International. Though Barnes never insists that exploring Vostok should be banned outright, ASOC is working through the International Union for the Conservation of Nature to sponsor a formal resolution that would halt drilling, going well beyond the protective intent of the Antarctic Treaty protocol. And if drilling proceeds anyway? "I seriously doubt that any Antarctic Treaty party would make such a move," contends Barnes. Though an ASOC resolution isn't binding, he says, the treaty's protocol is, and if a country violates it, environmental groups can file a lawsuit.

The scientists associated with Vostok don't share Roura's and

Barnes's views. For them, Vostok isn't just an abscess in the ice but a hole in our knowledge of the world — a peculiarly well-defined, romantically isolated, technologically dramatic hole. For good or ill, part of being a scientist is to find such holes irresistible. "Subglacial lakes are a brand-new environment," notes Ellis-Evans, "and, like anywhere humans come across, we are curious about it — we want to understand it. Whatever else, the place is a mystery. Scientists, being human, will always want to investigate a mystery. You cannot simply ignore it."

It's not that the scientists don't have doubts about what they're doing: The urge to protect and the urge to study still take turns leading their dance into the depths. If the lake is contaminated, measurements and samples will be compromised. But there's a deeper unease, too. "We have detected, explored, and spoiled almost all sites of the world, and we are going to do so with extraterrestrial systems," said Psenner in his Cambridge presentation. "Why not leave Lake Vostok in the mythical darkness?" But the nature of their profession is to overcome those doubts. As Psenner said at the meeting's end, "Vostok is possibly the only pristine lake we have left on Earth, and we should be extremely careful — but I'm too tempted by it not to go."

Put enough Vostok researchers in a room, and the lake will seem as hemmed in by different lines of questioning as it is by ice. But when it comes to Europa, there are really only two questions that count: Is there an ocean? And is anything swimming in it? In the pictures sent from *Voyager* probes in the 1970s and by *Galileo* during the past five years, the Jovian moon looks like an antique billiard ball — a smooth sphere of spreading cracks and stubborn stains. Europa's icy surface is indeed ancient by most measures, but by planetary standards it's freshly minted. A relative absence of impact craters suggests that the outer crust is only 100 million years old, in parts much less. Since Europa is a small world, half the diameter of Earth, this is quite surprising: The heat should have drained out of Europa's tiny core ages ago. But Jupiter's massive gravity alternately stretches and compresses Europa, and this kneading generates heat through friction. In fact, it may create enough heat to liquefy an entire ocean's worth of water under a layer of ice only a few miles thick. That makes it "a fascinating place

— and a prime place for space missions," says Frank Carsey, a team leader of the Polar Oceanography Group at JPL, which does ice research in Antarctica.

The idea of a habitable ocean locked beneath the permanent ice cap on Europa excites planetary scientists because on Earth, at least, all life requires liquid water. And beyond Earth, there's little direct evidence of any liquid water. Venus has baked itself into aridity. Mars was apparently wet once, but aside from its ice-capped poles, the Red Planet makes the Gobi desert look like a swamp. Comets and some asteroids contain ice in abundance — but if ice were what life needed, Antarctica would be green, not white. It's liquid water that counts, and Europa, with an ocean possibly greater in volume than all the oceans on Earth combined, offers a lot of it.

Every observation made with *Galileo's* instruments seems to argue for a Europan ocean. A follow-up probe, the still-unnamed Europa Orbiter, will be launched in 2003 to put the matter to rest. The craft will circle Europa from 124 miles above for nearly a month, measuring the tidal rises and falls of its surface with a laser altimeter. If the ice sits on rock, the tides will be small — solid ice would deform only slightly under Jupiter's insistent gravitational pull. But if there's water below the ice, the tidal bulge should stretch the moon's surface dramatically.

At many scientific centers — including the National Academy of Sciences, which keeps watch over NASA's science priorities — evidence of a Europan ocean would place it alongside Mars as the focus of the search for life elsewhere in the solar system. If the proposed Europa Orbiter confirms an ocean's existence — which most believe is 90 percent probable — a series of missions to Europa is likely to follow. "I'd bet there's life on Europa," says Richard Greenberg of the University of Arizona's Lunar and Planetary Laboratory in Tucson and one of the scientists analyzing imagery from *Galileo.* "I wouldn't bet there's life on Mars."

Finding life, either in Vostok or on Europa, requires drilling through an icy crust, which means resolving a multitude of engineering obstacles. While exploring a subsurface Antarctic lake is tricky, it doesn't compare with the logistics of controlling a probe in the solar system's farthest reaches. Horvath addressed some of those obstacles in 1996, when she asked her undergraduate students at Caltech and the United Kingdom's Leicester University to think about ways to get through Europa's crust to the putative ocean beneath. They soon realized that experts on earthly ice caps would have something useful to say about the matter. As it happens, JPL employs several climatologists, including Carsey, who had an interest in Antarctica. After their first meeting, he and Horvath resolved that Europa and Vostok were pieces of the same puzzle. "He says I thought of it; I say he thought of it," Horvath says with a laugh. The two of them attracted a clique of fellow Vostok enthusiasts who met regularly to discuss the kinds of hardware and strategies required to study life in dark waters under deep ice.

"The key technologies," Carsey explains, "are decontamination, then robotics, then instrumentation, then communications." In each area, the difficulties of drilling down to Vostok should illuminate the challenges of burrowing into Europa. There are, of course, major differences between the two projects. A probe going into Vostok can expect some hands-on help; a Europan probe will have to make its own decisions halfway across the solar system.

Carsey's cryobot offers a preview of what the probe might look like. He says it will be able to puncture a hole through the ice sheet with a hot-water drill, bathe itself in sterilizing chemicals, then cruise Vostok's deepest reaches. "It's probably not that hard to make a cryobot," says Carsey, "but to make one that's reliable and optimized — that's a little involved." If he can secure funding from the National Science Foundation, the first $100,000 prototype should be ready by summer. And in less than three years, he says, one of these icebots should be finished and ready for field testing.

During these trial runs, humans will be on hand to drill the cryobot's access hole and monitor its descent. The fuel needed to create boreholes in the ice will be flown in on C-130s by the New York Air National Guard, which supplies logistics support throughout the Antarctic. Such a cryobot would be employed for a Europa mission as well, which Carsey is considering as he puts together a blueprint for its design. But on Europa there would be no lifeline for handlers trying to send instantaneous commands to the bot far below: Signals sent to Europa will take thirty-five to fifty minutes to reach the probe in a place that makes Vostok station look like Mauritius. Europa's surface gets no warmer than minus 250 degrees, the sunlight is twenty-seven times dimmer than it is on Earth, there's no atmosphere to speak of, and the place is continuously bombarded by a vicious hail of particles from Jupiter's radiation belts.

Just getting to Europa is hard enough. The orbiter due to launch in 2003 will take almost three years to reach the moon, and radiation damage to its solar panels will make it reliable for only one month of work after it arrives. An energy-intensive project like drilling a hole through the ice will almost certainly require nuclear power — technology NASA has employed before, but which encounters increasing opposition. "Some environmentalists complain that the radiothermal generators are dangerous to launch, and are pollutive once they get to their target," says Carsey. "Both are arguably true. But the RTGs are designed to survive a launch calamity without breaking up — and, in a case like Europa's, the surface of the planet is far more radioactive than the RTG itself."

Lake Vostok's mystique is drawing people like Arthur Lonne Lane, a colleague of Horvath's and Carsey's at JPL who is taking what he's learned building spacecraft instruments and applying it to developing "technologies to solve some of the problems we'll face once we begin the scientific explorations of Lake Vostok and Europa." A device entering Vostok — and, someday, Europa's oceans — would have to be extremely compact, he says, small enough to be contained inside Carsey's cryobot. "The deep-water cameras they put on *Alvin* and other submersibles take beautiful pictures, but their housings are big and long," he says, going on to describe something about the size of a filing cabinet. "They have a huge glass dome at the front and these enormous outside arc lights."

Lane's instruments, by contrast, are economical: long, thin tubes of titanium built to be stuck into volcanic vents on the ocean floor. His latest camera — which has worked at depths of 1.2 miles — uses optical fibers to shine light from internal lasers out into the world, with a camera and spectrograph peering through a small, sturdy window at whatever they illuminate. "We developed a camera system that could operate at 400 degrees Celsius (752 degrees Fahrenheit), appropriate for hot vents and future deep-atmosphere probes of Jupiter, Venus, and Saturn," says Lane. "For Vostok and Europa, hot temperatures aren't anticipated, so the existing design wouldn't change a bit." In November, Lane will send a newer version of his imaging system nearly a half-mile into an Antarctic ice floe to study its base — a shakedown for future missions to Vostok and Europa.

Lane's camera would get to the bottom of an ice sheet with the help of Hermann Engelhardt, a glaciologist at Caltech. Engelhardt studies the conditions that lubricate ice sheets and glaciers as they move across terrain. This means he needs ways of looking beneath the ice, and his tool of choice is the hot-water drill — a vertically mounted, high-pressure fire hose. Hot water cuts through the ice ahead, and cooler water is pumped out of the hole from behind the drill head, then reheated and pumped back through the nozzle. In 1998, Engelhardt used this technology to drill forty-three holes, each through an ice sheet in a couple of months, leaving instruments underneath that he could then contact via satellite from his laboratory in Pasadena. Next year he'll lower one of Lane's probes into such a hole to see what's going on in far greater detail.

So when can we expect the first mission to get under way? After Carsey's cryobot — which incorporates Lane's camera and Engelhardt's drill — has been tested, after governments have agreed to pay, after environmental-impact assessments have satisfied the watchdogs at SCAR, and after further surveying has provided precise data about the lake's shape and the distribution of its sediments. When will that be, exactly? "I have a different answer every time somebody asks me," says Carsey. "What I really think is that we're probably about five years away."

But whether it's five years or twice that long, he sees the mission's progress like this: First, a hot-water drilling system creates a borehole down to within 500 yards of the lake's surface. "The approach is nice and fast," says Carsey. "It would maybe take a week to get to that depth." A long, thin cryobot, meticulously cleaned and packaged in a sterile wrapping, is lowered nose-first into the hole. "The working versions would be made of stainless steel or titanium," he says, "cylindrical in shape, probably ten to fifteen centimeters in diameter — small is better — and one to two meters long."

Once the cryobot is in position, the water above it is allowed to freeze, and a sterilizing solution is released. Electric current starts to flow down the 2.5 miles of cable — now sealed in ice — that link the cryobot to its controller. The probe's nose starts to heat up, a little of the surrounding ice starts to melt — slowly at first — and the cryobot starts to sink under its own weight. Pumps in its flanks suck up some of the melted water, which is reheated and pumped back out through the nose. This self-contained hot-water drill burrows downward, paying out fresh cable. As the probe sinks, the hole behind it freezes up. It comes to the lake's surface, pauses, and enters. Only the cryobot goes in; only information comes out.

The cryobot hangs still for a while just beneath the ice, its lasers the first source of light to illuminate the lake in millions of years, shining a few tens of yards into the darkness. The camera surveys the ice surface, searching for any clumps of matter and reading their spectral signatures. Digital images start to flow up through the ice, out to the satellites above and to waiting scientists all around the world. Next, the cryobot begins gathering water samples, running tests to reveal any microbial activity. "We want images and chemical and biological data," says Carsey, "with data-handling systems on the bot and on the controller."

Slowly, the probe pays out more of its lifeline. Its sensors alert for any slow currents — for any subtle distinctions between layers of water — it sinks below the lake's top layer of ice. The drilling pumps come back to life, nudging the probe gently from side to side. A pendulum in the dark, it swings back and forth, deeper and deeper. Eventually, it all but grazes the lakebed sediments, scanning for clues to an unseen ecosystem. Finally, its oscillation damped, camera rolling, it hovers motionless, while the sensors embedded in its flanks scoop up sediment samples, sniffing out any potential for the production of energy — a sign of the fires of life.

And after that? More sophisticated instrumentation, the return of sediment samples to the surface (a far more difficult proposition that Carsey refuses to think about on the grounds that he'll be retired by then), perhaps even free-swimming robots that could seek out signs of hot springs.

And what about Europa? Armed with Antarctic experience and the results from the Europa Orbiter, Carsey imagines the first generation of extraterrestrial cryobots could be in full development later this decade. By the '10s and '20s, exploration of the satellite's ice and water should begin in earnest, an effort that will dwarf the Vostok research both in cost and potential reward: evidence of life elsewhere in the solar system.

Caltech's Joe Kirschvink, the earth scientist who came up with the notion of the snowball Earth, thinks finding complex organisms in the Europan depths is a long shot. With no surface life to supply electron acceptors, the ocean wouldn't offer chemical imbalances strong enough to keep life going. Earth's snowball episodes were planetary near-death experiences. Europa's far-longer freeze would be terminal. Other scientists are more optimistic, seeing alternative chemical states for the ocean that might yield viable environments. The radiation that makes the moon's surface so inhospitable might turn out to be a blessing in that the chemical radicals it creates in the ice could somehow make their way to the ocean and provide another source of energy.

The University of Arizona's Richard Greenberg argues that people doubtful about life on Europa are ignoring evidence that the ice is very thin in some places, which means there could be water close enough to the surface for photosynthesis to occur. The lake's organisms would spend most of their time frozen and inert, springing into action only when the tides created new cracks to feed them water — or when a burrowing cryobot passed by.

Whether Europa harbors life or not, it will merit decades of exploration. Over that sort of time span, Vostok's comparatively meager supply of knowledge will be drained, and knowledge is the only thing Vostok has to offer: It's hardly a place where anyone might try to live. Mars could one day be colonized; eager scientists might push for a base on Europa if the natives there were interesting enough. In some distant century, we could even establish outposts on planets in other solar systems. But there's no reason for humans to venture in person to cold, delicate, joyless Vostok, so there's no reason — after scientists lose interest — that Vostok shouldn't be returned to its previous isolation, left just as we found it, emptied of knowledge yet still filled with that gin-clear water.

Restored to its mythical darkness, Vostok's isolated ecosystem could last a very long time — longer than a mountain range, perhaps longer than an ocean basin. The East Antarctic ice sheet would most likely remain stable, even under severe global warming. In fact, there's no compelling reason to think it will disappear, until the insulating ring of ocean currents that cut it off from the warmer world is interrupted by some new arrangement of the continents.

In time — when ice has covered Vostok for two, three, or four times longer than it has so far — tectonic forces will pull the frozen continent away from its wallflower existence at the pole, nudging it closer to the growing low-latitude supercontinent of Asia-Australia. The ocean currents will shift, the warmer weather of the north will start nibbling at the ice. As the heavy ice sheets begin to melt and slip away, the continent will rise up, its unburdened rocks suddenly buoyant. Hills once hidden beneath ice will rise to become mountains. Life will reappear: soil bacteria, lichens, insects, and flowers seeded by the wind. Birds will return — first to the sky, then to the softening soil. In the end, the sun's faint glow will reach down through the last remaining lens of ice. After the longest night, a soft, diffuse dawn will come to Vostok. The thinned ice will crack, and the waters will suck new breath from the sky. The 100-million-year voyage through the darkness will end. And if it was ever there — if it managed to survive, if we did our part to protect its uniqueness — an age-old ecosystem will become a new scrap of food for the creatures of a strange, changed world.

VAL PLUMWOOD

Being Prey

FROM *Utne Reader*

IN THE EARLY WET SEASON, Kakadu's paperbark wetlands are especially stunning, as the water lilies weave white, pink, and blue patterns of dreamlike beauty over the shining thunderclouds reflected in their still waters. Yesterday, the water lilies and the wonderful bird life had enticed me into a joyous afternoon's idyll as I ventured onto the East Alligator Lagoon for the first time in a canoe lent by the park service. "You can play about on the backwaters," the ranger had said, "but don't go onto the main river channel. The current's too swift, and if you get into trouble, there are the crocodiles. Lots of them along the river!" I followed his advice and glutted myself on the magical beauty and bird life of the lily lagoons, untroubled by crocodiles.

Today, I wanted to repeat that experience despite the drizzle beginning to fall as I neared the canoe launch site. I set off on a day trip in search of an Aboriginal rock art site across the lagoon and up a side channel. The drizzle turned to a warm rain within a few hours, and the magic was lost. The birds were invisible, the water lilies were sparser, and the lagoon seemed even a little menacing. I noticed now how low the fourteen-foot canoe sat in the water, just a few inches of fiberglass between me and the great saurians, close relatives of the ancient dinosaurs. Not long ago, saltwater crocodiles were considered endangered, as virtually all mature animals in Australia's north were shot by commercial hunters. But after a decade and more of protection, they are now the most plentiful of the large animals of Kakadu National Park. I was actively involved in preserving such places, and for me, the crocodile was a symbol of the power and integrity of this place and the incredible richness of its aquatic habitats.

After hours of searching the maze of shallow channels in the swamp, I had not found the clear channel leading to the rock art site, as shown on the ranger's sketch map. When I pulled my canoe over in driving rain to a rock outcrop for a hasty, sodden lunch, I experienced the unfamiliar sensation of being watched. Having never been one for timidity, in philosophy or in life, I decided, rather than return defeated to my sticky trailer, to explore a clear, deep channel closer to the river I had traveled along the previous day.

The rain and wind grew more severe, and several times I pulled over to tip water from the canoe. The channel soon developed steep mud banks and snags. Farther on, the channel opened up and was eventually blocked by a large sandy bar. I pushed the canoe toward the bank, looking around carefully before getting out in the shallows and pulling the canoe up. I would be safe from crocodiles in the canoe — I had been told — but swimming and standing or wading at the water's edge were dangerous. Edges are one of the crocodile's favorite food-capturing places. I saw nothing, but the feeling of unease that had been with me all day intensified.

The rain eased temporarily, and I crossed a sandbar to see more of this puzzling place. As I crested a gentle dune, I was shocked to glimpse the muddy waters of the East Alligator River gliding silently only 100 yards away. The channel had led me back to the main river. Nothing stirred along the riverbank, but a great tumble of escarpment cliffs up on the other side caught my attention. One especially striking rock formation — a single large rock balanced precariously on a much smaller one — held my gaze. As I looked, my whispering sense of unease turned into a shout of danger. The strange formation put me sharply in mind of two things: of the indigenous Gagadgu owners of Kakadu, whose advice about coming here I had not sought, and of the precariousness of my own life, of human lives. As a solitary specimen of a major prey species of the saltwater crocodile, I was standing in one of the most dangerous places on earth.

I turned back with a feeling of relief. I had not found the rock paintings, I rationalized, but it was too late to look for them. The strange rock formation presented itself instead as a telos of the day, and now I could go, home to trailer comfort.

As I pulled the canoe out into the main current, the rain and wind started up again. I had not gone more than five or ten minutes down the channel when, rounding a bend, I saw in midstream what looked like a floating stick — one I did not recall passing on my way up. As the current moved me toward it, the stick developed eyes. A crocodile! It did not look like a large one. I was close to it now but was not especially afraid; an encounter would add interest to the day.

Although I was paddling to miss the crocodile, our paths were strangely convergent. I knew it would be close but I was totally unprepared for the great blow when it struck the canoe. Again it struck, again and again, now from behind, shuddering the flimsy craft. As I paddled furiously, the blows continued. The unheard of was happening; the canoe was under attack! For the first time, it came to me fully that I was prey. I realized I had to get out of the canoe or risk being capsized.

The bank now presented a high steep face of slippery mud. The only obvious avenue of escape was a paperbark tree near the muddy bank wall. I made the split-second decision to leap into its lower branches and climb to safety. I steered to the tree and stood up to jump. At the same instant, the crocodile rushed up alongside the canoe, and its beautiful, flecked golden eyes looked straight into mine. Perhaps I could bluff it, drive it away, as I had read of British tiger hunters doing. I waved my arms and shouted, "Go away!" (We're British here.) The golden eyes glinted with interest. I tensed for the jump and leapt. Before my foot even tripped the first branch, I had a blurred, incredulous vision of great toothed jaws bursting from the water. Then I was seized between the legs in a red-hot pincer grip and whirled into the suffocating wet darkness.

Our final thoughts during near-death experiences can tell us much about our frameworks of subjectivity. A framework capable of sustaining action and purpose must, I think, view the world "from the inside," structured to sustain the concept of a continuing, narrative self; we remake the world in that way as our own, investing it with meaning, reconceiving it as sane, survivable, amenable to hope and resolution. The lack of fit between this subject-centered version and reality comes into play in extreme moments. In its final, frantic attempts to protect itself from the knowledge that threatens the narrative framework, the mind can instantaneously fabricate terminal doubt of extravagant proportions: *This is not really happening. This is a nightmare from which I will soon awake.* This desperate delusion split apart as I hit the water. In that flash, I glimpsed the world for the first time "from the outside," as a world no longer my own, an unrecognizable bleak landscape composed of raw necessity, indifferent to my life or death.

Few of those who have experienced the crocodile's death roll have lived to describe it. It is, essentially, an experience beyond words of total terror. The crocodile's breathing and heart metabolism are not suited to prolonged struggle, so the roll is an intense burst of power designed to overcome the victim's resistance quickly. The crocodile then holds the feebly struggling prey underwater until it drowns. The roll was a centrifuge of boiling blackness that lasted for an eternity, beyond endurance, but when I seemed all but finished, the rolling suddenly stopped. My feet touched bottom, my head broke the surface, and, coughing, I sucked at air, amazed to be alive. The crocodile still had me in its pincer grip between the legs. I had just begun to weep for the prospects of my mangled body when the crocodile pitched me suddenly into a second death roll.

When the whirling terror stopped again I surfaced again, still in the crocodile's grip next to a stout branch of a large sandpaper fig growing in the water. I grabbed the branch, vowing to let the crocodile tear me apart rather than throw me again into that spinning, suffocating hell. For the first time I realized that the crocodile was growling, as if angry. I braced myself for another roll, but then its jaws simply relaxed: I was free. I gripped the branch and pulled away, dodging around the back of the fig tree to avoid the forbidding mud bank, and tried once more to climb into the paperbark tree.

As in the repetition of a nightmare, the horror of my first escape attempt was repeated. As I leapt into the same branch, the crocodile seized me again, this time around the upper left thigh, and pulled me under. Like the others, the third death roll stopped, and we came up next to the sandpaper fig branch again. I was growing weaker, but I could see the crocodile taking a long time to kill me this way. I prayed for a quick finish and decided to provoke it by attacking it with my hands. Feeling back behind me along the head, I encountered two lumps. Thinking I had the eye sockets, I jabbed my thumbs into them with all my might. They slid into warm, unresisting holes (which may have been the ears, or perhaps the nostrils), and the crocodile did not so much as flinch. In despair, I grabbed the branch again. And once again, after a time, I felt the crocodile jaws relax, and I pulled free.

I knew I had to break the pattern; up the slippery mud bank was the only way. I scrabbled for a grip, then slid back toward the waiting jaws. The second time I almost made it before again sliding back, braking my slide by grabbing a tuft of grass. I hung there, exhausted. *I can't make it,* I thought. *It'll just have to come and get me.* The grass tuft began to give way. Flailing to keep from sliding farther, I jammed my fingers into the mud. This was the clue I needed to survive. I used this method and the last of my strength to climb up the bank and reach the top. I was alive!

Escaping the crocodile was not the end of my struggle to survive. I was alone, severely injured, and many miles from help. During the attack, the pain from the injuries had not fully registered. As I took my first urgent steps, I knew something was wrong with my leg. I did not wait to inspect the damage but took off away from the crocodile toward the ranger station.

After putting more distance between me and the crocodile. I stopped and realized for the first time how serious my wounds were. I did not remove my clothing to see the damage to the groin area inflicted by the first hold. What I could see was bad enough. The left thigh hung open, with bits of fat, tendon, and muscle showing, and a sick, numb feeling suffused my entire body. I tore up some clothing to bind the wounds and made a tourniquet for my bleeding thigh, then staggered on, still elated from my escape. I went some distance before realizing with a sinking heart that I had crossed the swamp above the ranger station in the canoe and could not get back without it.

I would have to hope for a search party, but I could maximize my chances by moving downstream toward the swamp edge, almost two miles away. I struggled on, through driving rain, shouting for mercy from the sky, apologizing to the angry crocodile, repenting to this place for my intrusion. I came to a flooded tributary and made a long upstream detour looking for a safe place to cross.

My considerable bush experience served me well, keeping me on course (navigating was second nature). After several hours, I began to black out and had to crawl the final distance to the swamp's edge. I lay there in the gathering dusk to await what would come. I did not expect a search party until the following day, and I doubted I could last the night.

The rain and wind stopped with the onset of darkness, and it grew perfectly still. Dingoes howled, and clouds of mosquitoes whined around my body. I hoped to pass out soon, but consciousness persisted. There were loud swirling noises in the water, and I knew I was easy meat for another crocodile. After what seemed like a long time, I heard the distant sound of a motor and saw a light moving on the swamp's far side. Thinking it was a boat, I rose up on my elbow and called for help. I thought I heard a faint reply, but then the motor grew fainter and the lights went away. I was as devastated as any castaway who signals desperately to a passing ship and is not seen.

The lights had not come from a boat. Passing my trailer, the ranger noticed there was no light inside it. He had driven to the canoe launch site on a motorized trike and realized I had not returned. He had heard my faint call for help, and after some time, a rescue craft appeared. As I began my thirteen-hour journey to Darwin Hospital, my rescuers discussed going upriver the next day to shoot a crocodile. I spoke strongly against this plan: I was the intruder, and no good purpose could be served by random revenge. The water around the spot where I had been lying was full of crocodiles. That spot was under six feet of water the next morning, flooded by the rains signaling the start of the wet season.

In the end I was found in time and survived against many odds. A similar combination of good fortune and human care enabled me to overcome a leg infection that threatened amputation or worse. I probably have Paddy Pallin's incredibly tough walking shorts to thank for the fact that the groin injuries were not as severe as the leg injuries. I am very lucky that I can still walk well and have lost few of my previous capacities. The wonder of being alive after being held — quite literally — in the jaws of death has never entirely left me. For the first year, the experience of existence as an unexpected blessing cast a golden glow over my life, despite the injuries and the pain. The glow has slowly faded, but some of that new gratitude for life endures, even if I remain unsure whom I should thank.

The gift of gratitude came from the searing flash of near-death knowledge, a glimpse "from the outside" of the alien, incomprehensible world in which the narrative of self has ended.

I had survived the crocodile attack, but not the cultural drive to represent it in terms of the masculinist monster myth: the master narrative. The encounter did not immediately present itself to me as a mythic struggle. I recall thinking with relief, as I struggled from the attack site, that I now had a good excuse for being late with an overdue article and a foolish but unusual story to tell a few friends. Crocodile attacks in North Queensland have often led to massive crocodile slaughters, and I feared that my experience might have put the creatures at risk again. That's why I tried to minimize publicity and save the story for my friends alone.

This proved to be extremely difficult. The media machine headlined a garbled version anyway, and I came under great pressure, especially from the hospital authorities, whose phone lines had been jammed for days, to give a press interview. We all want to pass on our story, of course, and I was no exception. During those incredible split seconds when the crocodile dragged me a second time from tree to water, I had a powerful vision of friends discussing my death with grief and puzzlement. The focus of my own regret was that they might think I had been taken while risking a swim. So important is the story and so deep the connection to others, carried through the narrative self, that it haunts even our final desperate moments.

By the same token, the narrative self is threatened when its story is taken over by others and given an alien meaning. This is what the mass media do in stereotyping and sensationalizing stories like mine — and when they digest and repackage the stories of indigenous peoples and other subordinated groups. As a story that evoked the monster myth, mine was especially subject to masculinist appropriation. The imposition of the master narrative occurred in several ways: in the exaggeration of the crocodile's size, in portraying the encounter as a heroic wrestling match, and especially in its sexualization. The events seemed to provide irresistible material for the pornographic imagination, which encouraged male identification with the crocodile and interpretation of the attack as sadistic rape.

Although I had survived in part because of my active struggle and bush experience, one of the major meanings imposed on my story was that the bush was no place for a woman. Much of the Australian media had trouble accepting that women could be competent in the bush, but the most advanced expression of this masculinist mindset was *Crocodile Dundee,* which was filmed in Kakadu not long after my encounter. Two recent escape accounts had both involved active women, one of whom had actually saved a man. The film's story line, however, split the experience along conventional gender lines, appropriating the active struggle and escape parts for the male hero and representing the passive "victim" parts in the character of an irrational and helpless woman who has to be rescued from the crocodile-sadist (the rival male) by the bushman hero.

I had to wait nearly a decade before I could repossess my story and write about it in my own terms. For our narrative selves, passing on our stories is crucial, a way to participate in and be empowered by culture. Retelling the story of a traumatic event can have tremendous healing power. During my recovery, it seemed as if each telling took part of the pain and distress of the memory away. Passing on the story can help us transcend not only social harm but also our own biological death. Cultures differ in how well they provide for passing on their stories. Because of its highly privatized sense of the individual, contemporary Western culture is, I think, relatively impoverished in this respect. In contrast, many Australian Aboriginal cultures offer rich opportunities for passing on stories. What's more, Aboriginal thinking about death sees animals, plants, and humans sharing a common life force. Their cultural stories often express continuity and fluidity between humans and other life that enables a degree of transcendence of the individual's death.

In Western thinking, in contrast, the human is set apart from nature as radically other. Religions like Christianity must then seek narrative continuity for the individual in the idea of an authentic self that belongs to an imperishable realm above the lower sphere of nature and animal life. The eternal soul is the real, enduring, and identifying part of the human self, while the body is animal and corrupting. But transcending death this way exacts a great price; it treats the earth as a lower, fallen realm, true human identity as outside nature, and it provides narrative continuity for the individual only in isolation from the cultural and ecological community and in opposition to a person's perishable body.

It seems to me that in the human supremacist culture of the West there is a strong effort to deny that we humans are also animals positioned in the food chain. This denial that we ourselves are food for others is reflected in many aspects of our death and burial practices — the strong coffin, conventionally buried well below the level of soil fauna activity, and the slab over the grave to prevent any other thing from digging us up, keeps the Western human body from becoming food for other species. Horror movies and stories also reflect this deep-seated dread of becoming food for other forms of life: Horror is the wormy corpse, vampires sucking blood, and alien monsters eating humans. Horror and outrage usually greet stories of other species eating humans. Even being nibbled by leeches, sandflies, and mosquitoes can stir various levels of hysteria.

This concept of human identity positions humans outside and above the food chain, not as part of the feast in a chain of reciprocity but as external manipulators and masters of it: Animals can be our food, but we can never be their food. The outrage we experience at the idea of a human being eaten is certainly not what we experience at the idea of animals as food. The idea of human prey threatens the dualistic vision of human mastery in which we humans manipulate nature from outside, as predators but never prey. We may daily consume other animals by the billions, but we ourselves cannot be food for worms and certainly not meat for crocodiles. This is one reason why we now treat so inhumanely the animals we make our food, for we cannot imagine ourselves similarly positioned as food. We act as if we live in a separate realm of culture in which we are never food, while other animals inhabit a different world of nature in which they are no more than food, and their lives can be utterly distorted in the service of this end.

Before the encounter, it was as if I saw the whole universe as framed by my own narrative, as though the two were joined perfectly and seamlessly together. As my own narrative and the larger story were ripped apart, I glimpsed a shockingly indifferent world in which I had no more significance than any other edible being. The thought, *This can't be happening to me. I'm a human being. I am more than just food!* was one component of my terminal incredulity.

It was a shocking reduction, from a complex human being to a mere piece of meat. Reflection has persuaded me that not just humans but any creature can make the same claim to be more than just food. We are edible, but we are also much more than edible. Respectful, ecological eating must recognize both of these things. I was a vegetarian at the time of my encounter with the crocodile, and remain one today. This is not because I think predation itself is demonic and impure, but because I object to the reduction of animal lives in factory farming systems that treat them as living meat.

Large predators like lions and crocodiles present an important test for us. An ecosystem's ability to support large predators is a mark of its ecological integrity. Crocodiles and other creatures that can take human life also present a test of our acceptance of our ecological identity. When they're allowed to live freely, these creatures indicate our preparedness to coexist with the otherness of the earth, and to recognize ourselves in mutual, ecological terms, as part of the food chain, eaten as well as eater.

Thus the story of the crocodile encounter now has, for me, a significance quite the opposite of that conveyed in the master/monster narrative. It is a humbling and cautionary tale about our relationship with the earth, about the need to acknowledge our own animality and ecological vulnerability. I learned many lessons from the event, one of which is to know better when to turn back and to be more open to the sorts of warnings I had ignored that day. As on the day itself, so even more to me now, the telos of these events lies in the strange rock formation, which symbolized so well the lessons about the vulnerability of humankind I had to learn, lessons largely lost to the technological culture that now dominates the earth. In my work as a philosopher, I see more and more reason to stress our failure to perceive this vulnerability, to realize how misguided we are to view ourselves as masters of a tamed and malleable nature. The balanced rock suggests a link between my personal insensitivity and that of my culture. Let us hope that it does not take a similar near-death experience to instruct us all in the wisdom of the balanced rock.

SANDRA POSTEL

Troubled Waters

FROM *The Sciences*

IN JUNE 1991, after a leisurely lunch in the fashionable Washington, D.C., neighborhood of Dupont Circle, Alexei Yablokov, then a Soviet parliamentarian, told me something shocking. Some years back he had had a map hanging on his office wall depicting Soviet central Asia without the vast Aral Sea. Cartographers had drawn it in the 1960s, when the Aral was still the world's fourth-largest inland body of water.

I felt for a moment like a cold war spy to whom a critical secret had just been revealed. The Aral Sea, as I knew well, was drying up. The existence of such a map implied that its ongoing destruction was no accident. Moscow's central planners had decided to sacrifice the sea, judging that the two rivers feeding it could be put to more valuable use irrigating cotton in the central Asian desert. Such a planned elimination of an ecosystem nearly the size of Ireland was surely one of humanity's more arrogant acts.

Four years later, when I traveled to the Aral Sea region, the Soviet Union was no more; the central Asian republics were now independent. But the legacy of Moscow's policies lived on: thirty-five years of siphoning the region's rivers had decreased the Aral's volume by nearly two thirds and its surface area by half. I stood on what had once been a seaside bluff outside the former port town of Muynak, but I could see no water. The sea was twenty-five miles away. A graveyard of ships lay before me, rotting and rusting in the dried-up seabed. Sixty thousand fishing jobs had vanished, and thousands of people had left the area. Many of those who remained suffered from a variety of cancers, respiratory ailments, and other diseases. Winds ripping across the desert were lifting tens of millions of tons of a toxic salt-dust chemical residue from the exposed seabed each year and dumping it on surrounding croplands and villages. Dust storms and polluted rivers made it hazardous to breathe the air and drink the water.

The tragedy of the Aral Sea is by no means unique. Around the world countless rivers, lakes, and wetlands are succumbing to dams, river diversions, rampant pollution, and other pressures. Collectively they underscore what is rapidly emerging as one of the greatest challenges facing humanity in the decades to come: how to satisfy the thirst of a world population pushing nine billion by the year 2050, while protecting the health of the aquatic environment that sustains all terrestrial life.

The problem, though daunting, is not insurmountable. A number of technologies and management practices are available that could substantially reduce the amount of water used by agriculture, industry, and households. But the sad reality is that the rules and policies that drive water-related decisions have not adequately promoted them. We have the ability to provide both people and ecosystems with the water they need for good health, but those goals need to be elevated on the political agenda.

Observed from space, our planet seems wealthy in water beyond measure. Yet most of the earth's vast blueness is ocean, far too salty to drink or to irrigate most crops. Only about 2.5 percent of all the water on earth is fresh water, and two thirds of that is locked away in glaciers and ice caps. A minuscule share of the world's water — less than one hundredth of 1 percent — is both drinkable and renewed each year through rainfall and other precipitation. And though that freshwater supply is renewable, it is also finite. The quantity available today is the same that was available when civilizations first arose thousands of years ago, and so the amount of water that should be allotted to each person has declined steadily with time. It has dropped by 58 percent since 1950, as the population climbed from 2.5 billion to 6 billion, and will fall an additional 33 percent within fifty years if our numbers reach 8.9 billion, the middle of the projected range.

Because rainfall and river flows are not distributed evenly throughout the year or across the continents, the task of adapting water to human use is not an easy one. Many rivers are tempestuous and erratic, running high when water is needed least and low when it is needed most. Every year two thirds of the water in the earth's rivers rushes untapped to the sea in floods. An additional one fifth flows in remote areas such as the Amazon basin and the Arctic tundra. In many developing countries monsoons bring between 70 and 80 percent of the year's rainfall in just three months, greatly complicating water management. When it comes to water, it seems, nature has dealt a difficult hand.

As a result, the history of water management has largely been one of striving to capture, control, and deliver water to cities and farms when and where they need it. Engineers have built massive canal networks to irrigate regions that are otherwise too dry to support the cultivation and growth of crops. The area of irrigated land worldwide has increased more than thirtyfold in the past two centuries, turning near-deserts such as southern California and Egypt into food baskets. Artificial oasis cities have bloomed. In Phoenix, Arizona, which gets about seven inches of rain a year, seemingly abundant water pours from taps. With a swimming pool, lawn, and an array of modern appliances, a Phoenix household can readily consume 700 gallons of water a day.

But while the affluent enjoy desert swimming pools, more than a billion of the world's people lack a safe supply of drinking water, and 2.8 billion do not have even minimal sanitation. The World Health Organization estimates that 250 million cases of water-related diseases such as cholera arise annually, resulting in between 5 and 10 million deaths. Intestinal worms infect some 1.5 billion people, killing as many as 100,000 a year. Outbreaks of parasitic diseases have sometimes followed the construction of large dams and irrigation systems, which create standing bodies of water where the parasites' hosts can breed. In sub-Saharan Africa, many women and girls walk several miles a day just to collect water for their families. Tens of millions of poor farm families cannot afford to irrigate their land, which lowers their crop productivity and leaves them vulnerable to droughts.

Even in countries in which water and sanitation are taken for granted, there are disturbing trends. Much of the earth's stable year-round water supply resides underground in geologic formations called aquifers. Some aquifers are nonrenewable — the bulk of their water accumulated thousands of years ago and they get little or no replenishment from precipitation today. And though most aquifers are replenished by rainwater seeping into the ground, in a number of the world's most important food-producing regions farmers are pumping water from aquifers faster than nature can replace it. Aquifers are overdrawn in several key regions of the United States, including California's Central Valley, which supplies half of the nation's fruits and vegetables, and the southern Great Plains, where grain and cotton farmers are steadily depleting the Ogallala, one of the planet's greatest aquifers.

The problem is particularly severe in India, where a national assessment commissioned in 1996 found that water tables in critical farming regions were dropping at an alarming rate, jeopardizing perhaps as much as one fourth of the country's grain harvest. In China's north plain, where 40 percent of that nation's food is grown, water tables are plunging by more than a meter a year across a wide area.

On the basis of the best available data, I estimate that global groundwater overpumping totals at least 160 billion cubic meters a year, an amount equal to the annual flow of two Nile Rivers. Because it takes roughly 1,000 cubic meters of water to produce one ton of grain, some 160 million tons of grain — nearly 10 percent of the global food supply — depend on the unsustainable practice of depleting groundwater. That raises an unsettling question: If humanity is operating under such an enormous deficit today, where are we going to find the additional water to satisfy future needs?

Another harbinger of trouble is that many major rivers now run dry for large parts of the year. Five of Asia's great rivers — the Indus and the Ganges in southern Asia, the Yellow in China, and the Amu Darya and Syr Darya in the Aral Sea basin — no longer reach the sea for months at a time. The Chinese call the Yellow River their mother river, reflecting its role as the cradle of Chinese civilization. Today the Yellow River supplies water to 140 million people and 18 million acres of farmland. Yet it has run dry in its lower reaches almost every year of this past decade, and the dry section often stretches nearly 400 miles upstream from the river's mouth. In 1997 the dry spell lasted a record 226 days.

Not surprisingly, as water becomes scarce, competition for it is intensifying. Cities are beginning to divert water from farms in north-central China, southern India, the Middle East, and the western United States. Moreover, the world's urban population is expected to double to 5 billion by 2025, which will further increase the pressure to shift water away from agriculture. How such a shift will affect food production, employment in rural areas, rural-to-urban migration, and social stability are critical questions that have hardly been asked, much less analyzed.

Competition for water is also building in international river basins: 261 of the world's rivers flow through two or more countries. In the vast majority of those cases there are no treaties governing how the river water should be shared. As demands tax the supply in those regions, tensions are mounting. In five water hot spots — the Aral Sea region, the Ganges, the Jordan, the Nile, and the Tigris-Euphrates — the population of the nations in each basin will probably increase by at least 30 percent and possibly by as much as 70 percent by 2025.

The plight of the Nile basin seems particularly worrisome. Last in line for Nile water, Egypt is almost entirely dependent on the river and currently uses two thirds of its annual flow. About 85 percent of the Nile's flow originates in Ethiopia, which to date has used little of that supply but is now constructing small dams to begin tapping the upper headwaters. Meanwhile, Egypt is pursuing two large irrigation projects that have put it on a collision course with Ethiopia. Although Nile-basin countries have been meeting regularly to discuss how they can share the river, no treaty that includes all the parties yet exists. Shortly after signing the historic peace accords with Israel in 1979, Egyptian president Anwar Sadat said that only water could make Egypt wage war again. He was referring not to another potential conflict with Israel but to the possibility of hostilities with Ethiopia over the Nile.

The story of the shrinking Aral Sea underscores another form of competition: the conflict between the use of water in agriculture and industry, on the one hand, and its ecological role as the basis of life and sustainer of ecosystem health, on the other. After I returned from the Aral Sea, I was tempted to view the sea and the communities around it as tragic victims of Communist central planning. A year later, however, in May 1996, I visited the delta of the Colorado River and found a depressingly similar story.

The Colorado delta had once been lush, supporting as many as 400 plant species and numerous birds, fish, and mammals. The great naturalist Aldo Leopold, who canoed through the delta in 1922, called it "a milk and honey wilderness," a land of "a hundred green lagoons." As I walked amid salt flats, mud-cracked earth, and murky pools, I could hardly believe I was in the same place that Leopold had described. The treaties that divide the Colorado River among seven U.S. states and Mexico had set aside nothing to protect the river system itself. More water was promised to the eight treaty parties than the river actually carries in an average year. As a result, the large dams and river diversions upstream now drain so much water that virtually nothing flows through the delta and out to the Gulf of California.

As in the Aral Sea basin, the Colorado predicament has caused more than an environmental tragedy. The Cocopa Indians have fished and farmed in the delta for more than 1,000 years. Now their culture faces extinction because too little river water makes it to the delta.

What was gained by despoiling such cultural and biological riches, by driving long-settled people from their homes and wildlife from its habitats? The answer seems to be more swimming pools in Los Angeles, more golf courses in Arizona, and more desert agriculture. To be sure, the tradeoff helped boost the U.S. gross national product, but at the untallied cost of irreplaceable natural and cultural diversity.

Given the challenges that lie ahead, how can the needs of an increasingly thirsty world be satisfied, without further destroying aquatic ecosystems? In my view, the solution hinges on three major components: allocating water to maintain the health of natural ecosystems, doubling the productivity of the water allocated to human activities, and extending access to a ready supply of water to the poor.

Just as people require a minimum amount of water to maintain good health, so do ecosystems — as the Aral Sea, the Colorado delta, and numerous other areas painfully demonstrate. As the human use of water nears the limits of the supply in many places, we must ensure the continued functioning of ecosystems and the invaluable services they perform. Providing that assurance will entail a major scientific initiative, aimed at determining safe limits of water usage from aquifers, rivers, lakes, and other aquatic systems.

Laws and regulations, guaranteeing continued health of those ecosystems, must also be put in place.

Australia and South Africa are now leading the way in such efforts. Officials in Australia's Murray-Darling River basin have placed a cap on water extractions — a bold move aimed at reversing the decline in the health of the aquatic environment. South Africa's new water laws call for water managers to allocate water for the protection of ecological functions as well as for human needs.

The United States is also making efforts to heal some of its damaged aquatic environments. A joint federal-state initiative is working to restore the health of California's San Francisco Bay delta, which is home to more than 120 species of fish and supports 80 percent of the state's commercial fisheries. In Florida an $8 billion federal-state project is attempting to repair the treasured Everglades, the famed "river of grass," which has shrunk in half in the past century alone. And across the country a number of dams are slated for removal in an effort to restore fisheries and other benefits of river systems.

The second essential component in meeting water needs for the future will be to maximize the use of every gallon we extract. Because agriculture accounts for 70 percent of the world's water usage, raising water productivity in farming regions is a top priority. The bad news is that today less than half the water removed from rivers and aquifers for irrigation actually benefits a crop. The good news is that there is substantial room for improvement.

Drip irrigation ranks near the top of measures that offer great untapped potential. A drip system is essentially a network of perforated plastic tubing, installed on or below the soil surface, that delivers water at low volumes directly to the roots of plants. The loss to evaporation or runoff is minimal. When drip irrigation is combined with the monitoring of soil moisture and other ways of assessing a crop's water needs, the system delivers 95 percent of its water to the plant, compared with between 50 and 70 percent for the more conventional flood or furrow irrigation systems.

Besides saving water, drip irrigation usually boosts crop yield and quality, simply because it enables the farmer to maintain a nearly ideal moisture environment for the plants. In countries as diverse as India, Israel, Jordan, Spain, and the United States, studies have consistently shown that drip irrigation not only cuts water use by between 30 and 70 percent, but also increases crop yields by between 20 and 90 percent. Those improvements are often enough to double the water productivity. Lands watered by drip irrigation now account for a little more than 1 percent of all irrigated land worldwide. The potential, however, is far greater.

The information revolution that is transforming so many facets of society also promises to play a vital role in transforming the efficiency of water use. The state of California operates a network of more than a hundred automated and computerized weather stations that collect local climate data, including solar radiation, wind speed, relative humidity, rainfall, and air and soil temperature, and then transmit the data to a central computer in Sacramento. For each remote site, the computer calculates an evapotranspiration rate, from which farmers can then calculate the rate at which their crops are consuming water. In that way they can determine, quite accurately, how much water to apply at any given time throughout the growing season.

As urban populations continue expanding in the decades ahead, household consumption of water will also need to be made more efficient. As part of the National Energy Policy Act, which was signed into law in late 1992, the United States now has federal water standards for basic household plumbing fixtures — toilets, faucets, and showerheads. The regulations require that manufacturers of the fixtures meet certain standards of efficiency — thereby building conservation into urban infrastructure. Water usage with those fixtures will be about a third less in 2025 than it would have been without the new standards. Similar laws could also help rapidly growing Third World cities stretch their scarce water supplies. One of the most obvious ways to raise water productivity is to use water more than once. The Israelis, for instance, reuse two thirds of their municipal wastewater for crop production. Because both municipal and agricultural wastewater can carry toxic substances, reuse must be carefully monitored. But by matching appropriate water quality to various kinds of use, much more benefit can be derived from the fresh water already under human control. And that implies that more can remain in its natural state.

The third component of the solution to water security for the future is perhaps also the greatest challenge: extending water and sanitation services to the poor. Ensuring safe drinking water is one of the surest ways to reduce disease and death in developing countries. Likewise, the most direct way of reducing hunger among the rural poor is to raise their productive capacities directly. Like trickle-down economics, trickle-down food security does not work well for the poor. Greater corn production in Iowa will not alleviate hunger among the poor in India or sub-Saharan Africa. With access to affordable irrigation, however, millions of poor farmers who have largely been bypassed by the modern irrigation age can raise their productivity and incomes directly, reducing hunger and poverty at the same time.

In many cases the problem is not that the poor cannot afford to pay for water but that they are paying unfair prices — often more than do residents of developed nations. It is not uncommon for poor families to spend more than a quarter of their income on water. Lacking piped-in water, many must buy from vendors who charge outrageous prices, often for poor-quality water.

In Istanbul, Turkey, for instance, vendors charge ten times the rate paid by those who enjoy publicly supplied water; in Bombay, the overcharge is a factor of twenty. A survey of households in Port-au-Prince, Haiti, found that people connected to the water system pay about a dollar per cubic meter ($3.78 per 1,000 gallons), whereas the unconnected must buy water from vendors for between $5.50 and $16.50 per cubic meter — about twenty times the price typically paid by urban residents in the United States.

Cost estimates for providing universal access to water and sanitation vary widely. But even the higher-end estimates — some $50 billion a year — amount to only 7 percent of global military expenditures. A relatively minor reordering of social priorities and investments — and a more comprehensive definition of security — could enable everyone to share the benefits of clean water and adequate sanitation.

Equally modest expenditures could improve the lot of poor farmers. In recent years, for instance, large areas of Bangladesh have been transformed by a human-powered device called a treadle pump. When I first saw the pump in action on a trip to Bangladesh in 1998, it reminded me of a StairMaster exercise machine, and it is operated in much the same way. The operator pedals up and down on two long poles, or treadles, each attached to a cylinder.

The upward stroke sucks shallow groundwater into one of the cylinders, while the downward stroke of the opposite pedal expels water from the other cylinder (that was sucked in on the preceding upward stroke) into a field channel.

The pump costs just thirty-five dollars, and with that purchase, farm families that previously were forced to let their land lie fallow during the dry season — and go hungry for part of the year — can grow an extra crop of rice and vegetables and take the surplus to market. Each pump irrigates about half an acre, which is appropriate for the small plots that poor farmers generally cultivate. The average net annual return on the investment has been more than $100 per pump, enabling families to recoup their outlay in less than a year.

So far Bangladeshi farmers have purchased 1.2 million treadle pumps, thereby raising the productivity of more than 600,000 acres of farmland and injecting an additional $325 million a year into the poorest parts of the Bangladeshi economy. A private-sector network of 70 manufacturers, 830 dealers, and 2,500 installers supports the technology, creating jobs and raising incomes in urban areas as well.

The treadle pump is just one of many examples of small-scale, affordable irrigation technologies that can help raise the productivity and the income of poor farm families. In areas with no perennial source of water, as in the drylands of south Asia and sub-Saharan Africa, a variety of so-called water-harvesting techniques hold promise for capturing and channeling more rainwater into the soil. In parts of India, for instance, some farmers collect rainwater from the monsoon season in earth-walled embankments, then drain the stored water during the dry season. The method, known as *haveli,* enables farmers to grow crops when their fields would otherwise be barren. Israeli investigators have found that another simple practice — covering the soil between rows of plants with polyethylene sheets — helps keep rainwater in the soil by cutting down on evaporation. The method has doubled the yields of some crops.

To avert much misery in this new century, the ways water is priced, supplied, and allocated must be changed. Large government subsidies for irrigation, an estimated $33 billion a year worldwide, keep prices artificially low — and so fail to penalize farmers for wasting water. Inflexible laws and regulations discourage the marketing of water, leading to inefficient distribution and use. Without rules to regulate groundwater extractions, the depletion of aquifers persists. And the failure to place a value on freshwater ecosystems — their role in maintaining water quality, controlling floods, and providing wildlife habitats — has left far too little water in natural systems.

Will we make the right choices in the coming age of water scarcity? Our actions must ultimately be guided by more than technology or economics. The fact that water is essential to life lends an ethical dimension to every decision we make about how it is used, managed, and distributed. We need new technologies, to be sure, but we also need a new ethic: All living things must get enough water before some get more than enough.

RICHARD PRESTON

The Genome Warrior

FROM *The New Yorker*

"CRAIG VENTER IS AN ASSHOLE. He's an idiot. He is a thorn in people's sides and an egomaniac," a senior scientist in the Human Genome Project said to me recently. The Human Genome Project is a nonprofit international research consortium that since the late 1980s has been working to decipher the complete sequence of nucleotides in human DNA. The human genome is the total amount of DNA that is spooled into a set of twenty-three chromosomes in the nucleus of every typical human cell. It is often referred to as the book of human life, and most scientists agree that deciphering it will be one of the great achievements of our time. The stakes, in money and glory, to say nothing of the future of medicine, are huge.

In the United States, most of the funds for the Human Genome Project come from the National Institutes of Health, and it is often referred to, in a kind of shorthand, as the "public project," to distinguish it from for-profit enterprises like the Celera Genomics Group, of which Craig Venter is the president and chief scientific officer. "In my perception," said the scientist who was giving me the dour view of Venter, "Craig has a personal vendetta against the National Institutes of Health. I look at Craig as being an extremely shallow person who is only interested in Craig Venter and in making money. Only God knows what those people at Celera are doing."

What Venter and his colleagues are doing is preparing to announce, in the next few days or weeks, that they have placed in the proper order something like 95 percent of the readable letters in the human genetic code. They refer to this milestone as First Assembly. They have already started selling information about the genome to subscribers. The Human Genome Project is also on the verge of announcing a milestone: what it calls a "working draft" of the genome, which is more than 90 percent complete and is available to anyone, free of charge, on a Web site called GenBank. It contains a large number of fragments that have not yet been placed in order, but scientists in the public project are scrambling to get a more complete assembly. Both images of the human genome — Celera's and the public project's — are becoming clearer and clearer. The human book of life is opening, and we hold it in our hands.

A human DNA molecule is about a meter long and a twenty-millionth of a meter wide — the width of twenty hydrogen atoms. It is shaped like a twisted ladder, and each rung of the ladder is made up of four nucleotides — adenine, thymine, cytosine, and guanine. The DNA code is expressed in combinations of the letters A, T, C, and G, the first letters of the names of the nucleotides. The human genome contains at least 3.2 billion letters of genetic code, about the number of letters in 2,000 copies of *Moby Dick.*

Perhaps 3 percent of the human code consists of genes, which hold recipes for making proteins. Human genes are stretches of between 1,000 and 1,500 letters of code, often broken into pieces and separated by long passages of DNA that don't code for protein. It is believed that there are somewhere between 30,000 and possibly more than 100,000 genes in the human genome (there's great puzzlement about the number). Much of the rest of the genome consists of blocks of seemingly meaningless letters, gobbledygook. These sections are referred to as junk DNA, although it may be that we just don't understand the function of the apparent junk.

The conventional route for announcing scientific breakthroughs is publication in a scientific journal, and both Celera and the public project plan to publish annotated versions of the human genome later this year, perhaps in *Science.* It is even possible that they will announce a collaboration and publish together. Although right now the two sides look like armies maneuvering for advantage, the leaders of the Human Genome Project have consistently denied that they are involved in some kind of competition.

"They're trying to say it's not a race, right?" Craig Venter said to me recently, in a shrugging sort of way. "But if two sailboats are sailing near each other, then by definition it's a race. If one boat wins, then the winner says, 'We smoked them,' and the loser says, 'We weren't racing — we were just cruising.'"

I first met Craig Venter on a windy day in summer nearly a year ago, at Celera's headquarters in Rockville, Maryland, a half-hour drive northwest of Washington, D.C. The company's offices and laboratories occupy a pair of five-story white buildings with mirrored windows, surrounded by beautiful groves of red oaks and tulip-poplar trees. One of the buildings contains rooms packed with row after row of DNA-sequencing machines of a type known as the ABI Prism 3700. The other building holds what is said to be the most powerful civilian computer array in the world; it is surpassed only, perhaps, by that of the Los Alamos National Laboratory, which is used for simulating nuclear bomb explosions. This second building also contains the Command Center, a room stuffed with control consoles and computer screens. People in the Command Center monitor the flow of DNA inside Celera. The DNA flows through the Prism machines twenty-four hours a day, seven days a week.

That day last summer, Venter moved restlessly around his office. There had been a spate of newspaper stories about the race to decode the complete genome, and about the pressure Celera was putting on its competitors. "We're scaring the shit out of everybody, including ourselves," he said to me. Venter is fifty-three years old, and he has an active, cherubic face on which a smile often flickers and plays. He is bald, with a fuzz of short hair at the temples, and his head is usually sunburned. He has bright blue eyes and a soft voice. He was wearing khaki slacks and a blue shirt, New Balance running shoes, a preppy tie with small turtles on it, and a Rolex watch. Venter's office looks into the trees, and that day leaves were spinning on branches outside the windows, flashing their white undersides and promising rain. Beyond the trees, a chronic traffic jam was occurring on the Rockville Pike. Celera is in an area along a stretch of Interstate 270 known as the Biotechnology Corridor, which is dense with companies specializing in the life sciences.

Celera Genomics is a part of the P.E. Corporation, which was called Perkin-Elmer before the company's chief executive, Tony L. White, split the business into two parts: P.E. Biosystems, which makes the Prism machine, and Celera. Venter owns 5 percent of Celera's stock, which trades, often violently, on the New York Stock Exchange. In recent months, the stock has been tossed by waves of panic selling and panic buying. Currently, the company is valued at $3 billion, more or less. At times, Craig Venter's net worth has slopped around by $100 million a day, like water going back and forth in a bathtub.

"Our fundamental business model is like Bloomberg's," Venter said. "We're selling information about the vast universe of molecular medicine." Venter believes, for example, that one day Celera will help analyze the genomes of millions of people as a regular part of its business — this will be done over the Internet, he says — and the company will then help design or select drugs tailored to patients' particular needs. Genomics is moving so fast that it is possible to think that in perhaps fifteen years you will be able to walk into a doctor's office and have your own genome interpreted. It could be stored in a smart card. (You would want to keep the card in your wallet, in case you landed in an emergency room.) Doctors would read the smart card, and it would show a patient's total biological-software code. They could see the bugs in the code, the genes that make you vulnerable to certain diseases. Everyone has bugs in his code, and knowing what they are will become a key to diagnosis and treatment. If you became sick, doctors could watch the activity of your genes, using so-called gene chips, which are small pieces of glass containing detectors for every gene. Doctors could track how the body was responding to treatment. All your genes could be observed, operating in an immense symphony.

Venter stopped moving briefly, and sat down in front of a screen and tapped a keyboard. A Yahoo! quote came up. "Hey, we're over twenty today," he said. (Celera's stock has since split. Adjusted for today's prices, it was trading at ten dollars a share; last week, it was trading at around seventy dollars a share.) I was standing in front of a large model of Venter's yacht, the *Sorcerer,* in which he won the 1997 Trans-Atlantic Race in an upset victory — it was the only major ocean race that Venter had ever entered. "I got the boat for a bargain from the guy who founded Lands' End," Venter said. "I like to buy castoff things on the cheap from ultra-rich people."

Venter went into the hallway, and I followed him. Celera was renovating its space, and tiles were hanging from the ceiling. Some had fallen to the floor. Black stains dripped out of air-conditioning vents, and sheets of plywood were lying around. Workmen were Sheetrocking walls, ripping up carpet, and installing light fixtures, and a smell of paint and spackle drifted in the air. We took the stairs to the basement and entered a room that held about fifty ABI Prism 3700 machines. Each Prism was the size of a small refrigerator and had cost $300,000. Prisms are the fastest DNA sequencers on earth. At the moment, they were reading the DNA of the fruit fly. This was a pilot project for the human genome. The machines contained lasers. Heat from the lasers seemed to ripple from the machines, even though they were being cooled by a circulation system that drew air through them. The lasers were shining light on tiny tubes through which strands of fruit-fly DNA were moving, and the light was passing through the DNA, and sensors were reading the letters of the code. Each machine had a computer screen on which blocks of numbers and letters were scrolling past. It was fly code.

"You're looking at the third-largest DNA-sequencing facility in the world," Venter said. "We also have the second-largest and the largest."

We got into an elevator. The walls of the elevator were dented and bashed. Venter led me into a vast, low-ceilinged room that looked out into the trees. This was the largest DNA-decoding factory on earth. The room contained 150 Prisms — $45 million worth of the machines — and more Prisms were due to be installed any day. Air ducts dangled on straps from the ceiling, and one wall consisted of gypsum board.

Venter moved restlessly through the unfinished space. "You know, this is the most futuristic manufacturing plant on the planet right now," he said. Outdoors, the rain came, splattering on the windows, and the poplar leaves shivered. We stopped and looked over a sea of machines. "You're seeing Henry Ford's first assembly plant," he said. "What don't you see? People, right? There are three people working in this room. A year ago, this work would have taken 1,000 to 2,000 scientists. With this technology, we are literally coming out of the dark ages of biology. As a civilization, we know far less than 1 percent of what will be known about biology, human physiology, and medicine. My view of biology is 'We don't know shit.'"

Celera's business model provokes some interesting questions, and some observers believe the company could fail. For instance, it appears to be burning through at least $150 million a year. But who will want to buy the information the company is generating, and how much will they pay for it? "There will be an incredible demand for genomic information," Venter assured me. "When the first electric-power companies strung up wires on power poles, there were a lot of skeptics. They said, 'Who's going to buy all that electricity?' We already have more than a hundred million dollars in committed subscription revenues over five years from companies that are buying genomic information from us — Amgen, Novartis, Pharmacia & Upjohn, and others. After we finish the human genome, we'll do the mouse, rice, rat, dog, cow, corn, maybe apple trees, maybe clover. We'll do the chimpanzee."

One day at Celera's headquarters, I was talking with a molecular biologist named Hamilton O. Smith, who won a Nobel Prize in 1978 as the codiscoverer of restriction enzymes, which are used to cut DNA in specific places. Scientists use the enzymes like scissors, chopping up pieces of DNA so that they can be studied or recombined with the DNA of other organisms. Without the means to do this, there would be no such thing as genetic engineering.

Ham Smith is in his late sixties. He is six feet five inches tall, with a shock of stiff white hair and a modest manner. "Have you ever seen human DNA?" he asked me, as he poked around his lab.

"No."

"It's beautiful stuff."

A box that held four small plastic tubes, each the size of a pencil stub, sat on a countertop. "These four tubes hold enough human DNA to do the entire human genome project," Smith said. "There's a couple of drops of liquid in each tube."

He held up one of the tubes and turned it over in the light to show me. A droplet of clear liquid moved back and forth. It was the size of a dewdrop. Then he held up a glass vial, and rocked it back and forth, and a crystal-clear, syrupy liquid oozed around in it. He explained that this was DNA he'd extracted from human blood — from white cells. "That's long, unbroken DNA. This liquid looks glassy and clear, but it's snotty. It's like sugar syrup. It really is a sugar syrup, because there are sugars in the backbone of the DNA molecule."

Smith picked up a pipette — a handheld device with a hollow plastic needle in it, which is used for moving tiny quantities of liquid from one place to another. His hands are large, but they moved with precision. Holding the pipette, he sucked up a droplet of DNA mixed with a type of purified salt water called buffer. He held the drop in the pipette for a moment, then let it go. The droplet drooled. It reminded me of a spider dropping down a silk thread.

"There the DNA goes; it's stringing," he said. "The pure stuff is gorgeous." The molecules were sliding along one another, like spaghetti falling out of a pot, causing the water to string out. "It's absolutely glassy clear, without color," he said. "Sometimes it pulls back into the tube and won't come out. I guess that's like snot, too, and then you have to almost cut it with scissors. The molecule is actually quite stiff. It's like a plumber's snake. It bends, but only so much, and then it breaks. It's brittle. You can break it just by stirring it."

The samples of DNA that Celera is using are kept in a freezer near Smith's office. When he wants to get some human DNA, he removes a vial of frozen white blood cells or sperm from the freezer. The vials have coded labels. He thaws the sample of cells or sperm, then mixes the material with salt water, along with a little bit of detergent. A typical human cell looks like a fried egg, and the nucleus of the cell resembles the yolk. The detergent pops the eggs and the yolks, and strands of DNA spill out in the salt water. The debris falls to the bottom, leaving tangles of DNA suspended in the liquid.

One of Smith's research associates, a woman named Cindi Pfannkoch, showed me what shattered DNA was like. Using a pipette, she drew a tiny amount of liquid from a tube and let a drop go on a sheet of wax, where it beaded up like a tiny jewel, the size of the dot over this "i." An ant could have drunk it in full.

"There are two hundred million fragments of human DNA in this drop," she said. "We call that a DNA library."

She opened a plastic bottle, revealing a white fluff. "Here's some dried DNA." She took up a pair of tweezers and dragged out some of the fluff. It was a wad of dried DNA from the thymus gland of a calf; the wad was about the size of a cotton ball, and it contained several million miles of DNA.

"In theory," Ham Smith said, "you could rebuild the entire calf from any bit of that fluff."

I placed some of the DNA on the ends of my fingers and rubbed them together. The stuff was sticky. It began to dissolve on my skin. "It's melting — like cotton candy," I said.

"Sure. That's the sugar in DNA," Smith said.

"Would it taste sweet?"

"No. DNA is an acid, and it's got salts in it. Actually, I've never tasted it."

Later, I got some dried calf DNA. I placed a bit of the fluff on my tongue. It melted into a gluey ooze that stuck to the roof of my mouth in a blob. The blob felt slippery on my tongue, and the taste of pure DNA appeared. It had a soft taste, unsweet, rather bland, with a touch of acid and a hint of salt. Perhaps like the earth's primordial sea. It faded away.

DNA from six donors who contributed their blood or semen was used for Celera's human genome project. The donors included both men and women, and a variety of ethnic groups. Just one person, a man, supplied the DNA for First Assembly. Only Craig Venter and one other person at the company are said to know who the donors are. "I don't know who they are, but I wouldn't be surprised if one of them is Craig," Ham Smith remarked.

Craig Venter grew up in a working-class neighborhood on the east side of Millbrae, on the San Francisco peninsula. His family's house was near the railroad tracks. One of his favorite childhood activities, he says, was to play chicken on the tracks. In high school, he excelled in science and shop. He built two speed boats, and spent a lot of time surfing Half Moon Bay. He attended two junior colleges in a desultory way, but mostly he surfed, until he enlisted in the navy. He had long blond hair and a crisp body then. He was a medical corpsman in Vietnam, and twice he was sentenced to the brig for disobeying orders.

Venter has a history of confrontation with government authorities. He told me that as an enlisted man in San Diego he was court-martialed for refusing a direct order given by an officer. "She happened to be a woman I was dating," Venter said. "We had a spat, and she ordered me to cut my hair. I refused." A friend of his, Ron Nadel, who was a doctor in Vietnam, recalls that one of Venter's blowups with authority involved "telling a superior officer to do something that was anatomically impossible." Venter worked for a year in the intensive-care ward at Da Nang hospital, where, he calculates, more than a thousand Vietnamese and American soldiers died during his shifts, many of them while the 1968 Tet offensive was going on. When he returned to the United States, Venter finished college and then earned a Ph.D. in physiology and pharmacology from the University of California at San Diego.

Venter is married to a molecular biologist, Claire Fraser, who is the president of The Institute for Genomic Research (TIGR, pronounced "Tiger"), in Rockville, a nonprofit institute that he and Fraser helped establish in 1992. In 1998, he endowed TIGR with half of his original stake in Celera — 5 percent of the company. The gift is currently worth about $150 million, and it will be used to analyze the genomes of microbes that cause malaria and cholera and other diseases.

A few years ago, Venter developed a hole in his intestine, due to diverticulitis. He collapsed after giving a speech, and nearly died. He is fine now, but he blames stress caused by his enemies for his burst intestine. Venter has enemies of the first water. They are brilliant, famous, articulate, and regularly angry. At times, Venter seems to thrive on his enemies' indignation with an indifferent grace, like a surfer shooting a tubular wave, letting himself be propelled through their cresting wrath. At other times, he seems baffled, and says he can't understand why they don't like him.

One of Venter's most venerable enemies is James Watson, who, with Francis Crick and Maurice Wilkins, won the Nobel Prize in Medicine in 1962 for discovering the shape of the DNA molecule — what they called the double helix. Watson helped found the Human Genome Project, and he was the first head of the NIH genome program. I visited him in his office at the Cold Spring Harbor Laboratory, on Long Island. The office is paneled in blond oak and has a magnificent eastward view across Cold Spring Harbor. Watson is now in his seventies. He has a narrow face, lopsided teeth, a frizz of white hair, sharp, restless eyes, a squint, and a dreamy way of speaking in sentences that trail off. He put his hands on his head and squinted at me. "In 1953 with our first paper on DNA, we never saw the possibility . . ." he said. He looked away, up at the walls. "No chemist ever thought we could read the molecule." But a number of biologists began to think that reading the human genome might just be possible, and by the mid-1980s Watson had become convinced that the decryption of the genome was an important goal and should he pursued, even if it cost billions and took decades.

Watson appeared before Congress in May of 1987 and asked for an initial annual budget of $30 million for the project. The original plan was to sequence the human genome by 2005, at a projected cost of about $3 billion. The principal work of the project is now carried out by five major DNA-sequencing centers, as well as by a number of smaller centers around the world — all academic, nonprofit labs. The big centers include one at Baylor University in Texas, one at Washington University in St. Louis, the Whitehead Institute at MIT, the Joint Genome Institute of the Department of Energy, and the Sanger Centre, near Cambridge, England. The Wellcome Trust of Great Britain — the largest nonprofit medical-research foundation in the world — is funding the Sanger work, which is to sequence a third of the human genome. One of the founding principles of the Human Genome Project was the immediate release of all the human code that was found, making it available free of charge and without any restrictions on who could use it or what anyone could do with it.

In 1984, Craig Venter had begun working at the NIH, where he eventually developed an unorthodox strategy for decoding bits of genes. At the time, other scientists were painstakingly reading the complete sequence of each gene they studied. This process seemed frustratingly slow to Venter. He began isolating what are called expressed sequence tags, or ESTs, which are fragments of DNA at the ends of genes. When the ESTs were isolated, they could be used to identify genes in a rough way. With the help of a few sequencing machines, Venter identified bits of thousands of human genes. This was a source of unease at the NIH, because it was a kind of skimming rather than a complete reading of genes. Venter published his method in 1991 in an article in *Science*, along with partial sequences from about 350 human genes. The method was not received well by many genome scientists. It was fast, easy, and powerful, but it didn't look elegant, and some scientists seemed threatened by it. Venter claims that two of his colleagues, who are now heads of public genome centers, asked him not to publish his method or move forward with it for fear they would lose their funding for genome sequencing.

The NIH decided to apply for patents on the gene fragments that Venter had identified. James Watson blew his stack over the idea of anyone trying to patent bits of genes, and he got into a hostile situation with the director of the NIH, Bernadine Healy, who defended the patenting effort. In July of 1991, during a meeting in Washington called by Senator Pete Domenici, of New Mexico, to review the genome program, Watson dissed Venter's methods. "It isn't science," he said, adding that the machines "could be run by monkeys."

It was a strange moment. The Senate hearing room was almost empty — few politicians were interested in genes then. But Craig Venter was sitting in the room. "Jim Watson was clearly referring to Craig as a monkey in front of a U.S. senator," another scientist who was there said to me. "He portrayed Craig as the village idiot of genomics." Venter seemed to almost thrash in his chair, stung by Watson's words. "Watson was the ideal father figure of genomics," Venter says. "And he was attacking me in the Senate, when I was relatively young and new in the field."

Today, James Watson insists that he wasn't comparing Craig Venter to a monkey. "It's the patenting of genes I was objecting to. That's why I used the word 'monkey'! I hate it!" he said to me. The patent office turned down the NIH's application, but a few years later, two genomics companies, Incyte and Human Genome Sciences, adopted the EST method for finding genes, and it became the foundation of their businesses — currently worth, combined, about $7 billion on the stock market. Incyte and Human Genome Sciences are Celera's main business competitors. Samuel Broder, the chief medical officer at Celera, who is a former director of the National Cancer Institute, said to me, heatedly, "None of the people who severely and acrimoniously criticized Craig for his EST method ever said they were personally sorry. They ostracized Craig and then went on to use his method with never an acknowledgment."

James Watson now says, "The EST method has proved immensely useful, and it should have been encouraged."

Venter was increasingly unhappy at the NIH. He had received a $10 million grant to sequence human DNA, and he asked for permission to use some of the money to do EST sequencing, but his request was denied by the genome project. Venter returned the grant money with what he says was a scathing letter to Watson. In addition, Claire Fraser had been denied tenure at the NIH. Her review committee (which was composed entirely of middle-aged men) explained to her that it could not evaluate her work independently of her husband's. At the time, Fraser and Venter had separate labs and separate research programs. Fraser considered suing the NIH for sex discrimination.

Watson was forced to resign as head of the genome project in April of 1992, in part because of the dispute over patenting Craig Venter's work. That summer, Venter was approached by a venture capitalist named Wallace Steinberg, who wanted to set up a company that would use Venter's EST method to discover genes, create new drugs, and make money. "I didn't want to run a company; I wanted to keep doing basic research," Venter says. But Steinberg offered Venter a research budget of $70 million over ten years — a huge amount of money, then, for biotech. Venter, along with Claire Fraser and a number of colleagues, left the NIH and founded TIGR, which is a nonprofit organization. At the same time, Steinberg established a for-profit company, Human Genome Sciences, to exploit and commercialize the work of TIGR, which was required to license its discoveries exclusively to its sister company. Thus Venter got millions of dollars for research, but he had to hand his discoveries over to Human Genome Sciences for commercial development. Venter had one foot in the world of pure science and one foot in a bucket of money.

By 1994, the Human Genome Project was mapping the genomes of model organisms, which included the fruit fly, the roundworm, yeast, and *E. coli* (the organism that lives in the human gut), but no genome of any organism had been completed, except virus genomes, which are relatively small. Venter and Hamilton O. Smith (who was then at the Johns Hopkins School of Medicine) proposed speeding things up by using a technique known as whole-genome shotgun sequencing. In shotgunning, the genome is broken into small, random, overlapping pieces, and each piece is sequenced, or read. Then the jumble of pieces is reassembled in a computer that compares each piece to every other piece and matches the overlaps, thus assembling the whole genome.

Venter and Smith applied for a grant from the NIH to shotgun-sequence the genome of a disease-causing bacterium called *H. Influenzae*, or *H. flu* for short. It causes fatal meningitis in children. They proposed to do it in just a year. *H. flu* has 1.8 million letters of code, which seemed massive then (though the human code is 2,000 times as long). The review panel at the NIH gave Venter's proposal a low score, essentially rejecting it. According to Venter, the panel claimed that an attempt to shotgun-sequence a whole microbe was excessively risky and perhaps impossible. He appealed. The appeals process dragged on, and he went about shotgunning *H. flu* anyway. Venter and the TIGR team had nearly finished sequencing the *H. flu* genome when, in early 1995, a letter arrived at TIGR saying that the appeals committee had denied the grant on the ground that the experiment wasn't feasible. Venter published the *H. flu* genome a few months later in *Science*. Whole-genome shotgunning had worked. This was the first completed genome of a free-living organism.

It seems quite possible that Venter's grant was denied because of politics. The review panel seems to have hated the idea of giving NIH money to TIGR to make discoveries that would be turned over to a corporation, Human Genome Sciences. It turned down the grant, in spite of the fact that "all the smart people knew the method was straightforward and would work," Eric Lander, the head of the genome center at MIT and one of the leaders of the public project, said to me.

Around this time, Wallace Steinberg died of a heart attack, and his death provided a catalyst for a split between TIGR and Human Genome Sciences, which was run by a former AIDS researcher, William Haseltine. Venter and Haseltine were widely known to dislike each other. Venter sold his stock in Human Genome Sciences because of the rift between them, and after Steinberg died the relationship between the two organizations was formally ended.

Late in 1997, TIGR was doing some DNA sequencing for the Human Genome Project, and Venter began going to some of the project's meetings. That was when he started calling the heads of the public project's DNA-sequencing centers the Liars' Club, claiming that their predictions about when they would finish a task and how much it would cost were false. This did not win Venter many friends. But he seemed to have a point.

Francis Collins, a distinguished medical geneticist from the University of Michigan, had become the head of the NIH genome program shortly after James Watson resigned in 1992. In early January

1998, an internal budget projection from Collins's office somehow found its way to Watson (he seems to find out everything that's happening in molecular biology). This budget projection — it is not clear whether it was formal or was just an unofficial projection — was a document about eight pages long. It contained a graph marked "Confidential" indicating that Collins planned to spend only $60 million per year on direct human-DNA sequencing through 2005. It also predicted that by that year — when the human genome was supposed to be completed — only 1.6 billion to 1.9 billion letters of human code would be sequenced, that is, slightly more than half of the human genome.

This upset Watson, and he decided to discuss it with Eric Lander. On January 17, Watson traveled to Rockefeller University, on the East Side of Manhattan, where Lander was giving the prestigious Harvey Lecture. The two men met after the lecture at the faculty club at Rockefeller. They were dressed in black tie and were somewhat inebriated. Traditionally among medical people, the Harvey Lecture is given and listened to under the influence.

The Rockefeller faculty club overlooks a lawn and sycamore trees and the traffic of York Avenue. Watson and Lander sat down with cognacs at a small table in a dim corner of the room, on the far side of a pool table, where they could talk without being overheard. Also present and drinking cognac was a biologist named Norton Zinder, who is one of Watson's best friends. Zinder, like Watson, is a founder of the Human Genome Project. One of the older men brought up the confidential budget document with Lander, and both of them began to press him about it. They felt that it provided evidence that Collins did not intend to spend more than $60 million a year on human-DNA sequencing — nowhere near enough to get the job done, they felt.

Watson evidently felt that Lander had influence with Francis Collins, and he urged him to try to persuade Collins to spend more on direct sequencing of human DNA, and to twist Congress's arm for more money.

Norton Zinder was somewhat impaired with cocktails. "This thing is potchkeeing along, going nowhere!" he said, hammering the little table and waving his arms as he spoke. For him, the issue was simple: he had had a quadruple coronary bypass, and he had been receiving treatments for cancer, and now he was afraid he would not live to see the deciphering of the human genome. This was intolerable. The human genome had begun to seem like a vision of Canaan to Norton Zinder, and he thought he wouldn't make it there.

Eric Lander did not view things the way the older biologists did. In his opinion, the problem was organizational. The Human Genome Project was "too bloody complicated, with too many groups." He felt the real problem was a lack of focus. He wanted the project to create a small, elite group that would do the major sequencing of human DNA — shock cavalry that would lead a charge into the human genome. Implicitly, he thought its leader should be Eric Lander.

The three men downed their cognacs with a sense of frustration. "I had essentially given up seeing the human genome in my lifetime," Zinder says.

At about the same moment that Watson and his friends were lamenting the slowness of the public project, the Perkin-Elmer Corporation, which was a manufacturer of lab instruments, was secretly talking about a corporate reorganization. It controlled more than 90 percent of the market for DNA-sequencing equipment, and it was developing the ABI Prism 3700. The Prism was then only a prototype sitting in pieces in a laboratory in Foster City, California, but already it looked as if it were going to be at least ten times faster than any other DNA-sequencing machine. Perkin-Elmer executives began to wonder just what it could do. One day Michael Hunkapillar, who was then the head of the company's instrument division, got out a pocket calculator and estimated that several hundred Prisms could whip through a molecule of human DNA in a few weeks, although only in a rough way. To fill in the gaps — places where the DNA code came out garbled or wasn't read properly by the machines — it would be necessary to sequence the molecule again and again. This is known as repeat sequencing, or manyfold coverage, and might take a few years. Hunkapillar persuaded the chief executive of Perkin-Elmer, Tony White, to restructure the business and create a genomics company.

In December 1997, executives from Perkin-Elmer began telephoning Venter to see if he'd be interested in running the new company. He blew them off at first, but in early February 1998, he went to California with a colleague, Mark Adams, to look at the prototype Prism. When they saw it, they immediately understood its significance. Before the end of that day, Venter, Adams, and Hunkapillar had laid out a plan for decoding the human genome. A month later, Norton Zinder, Watson's friend, flew to California to see the machine. "It was just a piece of equipment sitting on a table, but I said, 'That's it! We've got the genome!'" he recalled. Zinder joined Celera as a member of its board of advisers, and received stock in the company, which has considerably enriched him. ("The chemists have been cleaning up," he said to me. "Now biologists have their hands on the money, too.") Zinder and Watson have maintained their friendship but have agreed not to speak about Celera with each other. They evidently fear that one or both of them could have a stroke arguing about Craig Venter.

At eleven o'clock in the morning on May 8, 1998, Craig Venter and Mike Hunkapillar walked into the office of Harold Varmus, who was then the director of the NIH, and announced the pending formation of a corporation, led by Venter, that was going to decode the human genome. (Celera did not yet have a name.) They proposed to Varmus that the company and the public project collaborate, sharing their data and — this point is enormously important to scientists — sharing the publication of the human genome, which meant sharing the credit and the glory for having done the work, including the unspoken possibility of a Nobel Prize. Varmus strongly suspected that this wasn't a sincere offer, and he told them that he needed time, particularly to check with Francis Collins. Later that same day, Venter and Hunkapillar drove to Dulles Airport, where they met Collins at the United Airlines Red Carpet Club, and again offered collaboration. Venter recalls that Collins seemed upset. Collins recalls that he merely asked Venter for time to consider the offer. Time was one thing Venter was not prepared to give.

Venter had alerted the *New York Times* to the story about the creation of the new company, and just an hour or so after the meeting with Collins he called the *Times* and told the paper it should run it. In the story, Venter announced that he would sequence the human genome by 2001 — four years ahead of the public project — and he would do it, he claimed, for between $150 and $200 million — less than a tenth of the projected cost of the public project. The *Times* reporter, Nicholas Wade, implied that the Human Genome Project might not meet its goals and might be superfluous.

Four days later, on May 12, Venter and Hunkapillar went to the Cold Spring Harbor Laboratory, where a meeting of the heads of the Human Genome Project was taking place. Venter got up and told them, in effect, that they could stop working, since he was going to sequence the human genome *tout de suite.* Later that week, sitting beside Varmus and Collins at a press conference, Venter looked out at a room full of reporters and suggested that biology and society would be better off if the Human Genome Project shifted gears and moved forward to do the genome of the . . . mouse.

It was a fart in church of magnitude nine. "The mouse is essential for interpreting the human genome," Venter tried to explain, but that didn't help. In the words of one head of a sequencing center who was at the Cold Spring Harbor meeting, "Craig has a certain lack of social skills. He goes into that meeting thinking everyone is going to thank him for doing the human genome himself. The thing blew up into a huge explosion." The head of another center recalled, "Craig came up to me afterward, and he said, 'Ha, ha, I'm going to do the human genome. You should go do the mouse.' I said to him, 'You bastard. You *bastard*' and I almost slugged him."

They felt that Venter was trying to stake out the human genome for himself as a financial asset while at the same time stealing the scientific credit. They felt that he was belittling their work. Venter said that he would make the genome available to the public but would charge customers who wanted to see and work with Celera's analyzed data.

James Watson was furious. He did not like the idea of having to pay money to Craig Venter for anything. Watson did not attend Venter's presentation, but he appeared in the lobby afterward, where he repeatedly said to people, "He's Hitler. This should not be Munich." To Francis Collins he said, "Are you going to be Churchill or Chamberlain?"

Venter left the meeting soon afterward, and he and Watson have exchanged only chilly greetings since.

The British leaders of the public project — John Sulston, the director of the Sanger Centre, and Michael Morgan, of the Wellcome Trust — reacted swiftly to Venter's announcement. They were in England, but they flew to the United States, and the next day arrived at Cold Spring Harbor, where they found things in disarray, if not in fibrillation, with scientists wondering if the Human Genome Project was about to die. To a standing ovation, Michael Morgan got up and read a Churchillian statement declaring that the Wellcome Trust would nearly double its funding for the public project, and would decode a full third of the human genome, and would challenge any "opportunistic" patents of the genome. "We were reacting, in part, to Craig's suggestion that we just close up shop and go home," Morgan says now.

Venter also announced that Celera would use the whole-genome shotgun method. The public project was using a more conventional method. John Sulston and Robert Waterston, the head of the sequencing center at Washington University, published a letter in *Science* asserting that Venter's method would be "woefully inadequate." Francis Collins was quoted in *USA Today* as saying that Celera was going to produce "the Cliffs Notes or the *Mad Magazine* version" of the human genome. Collins says now that his words were taken out of context, and he regrets the quote.

The company forged from Perkin-Elmer amid the turmoil was the P.E. Corporation, which holds the P.E. Biosystems Group, the unit that makes the Prism machines, and Celera Genomics. Michael Hunkapillar, who is now the president of P.E. Biosystems, believed that he could sell a lot of machines to everyone, including the Human Genome Project. There was a fat profit margin in the chemicals the machines use. The chemicals cost far more than the machine over the machine's lifetime. This was the razor-blade principle: if you put razors in people's hands, you will make money selling blades.

In August, Incyte Pharmaceuticals announced that it was starting a human genome project of its own. In September, James Watson quietly went to some key members of Congress and persuaded them to spend more money on the public project. At the same time, the leaders of the project announced a radical new game plan: they would produce a "working draft" of the human genome by 2001 — a year ahead of when Venter said he'd be done — and a finished, complete version by 2003. An epic race had begun.

A couple of months ago, Michael Morgan, of the Wellcome Trust, was talking to me about Venter and what had happened with the creation of Celera. "From the first press release, Craig saw the public program as something he wanted to denigrate," Morgan said. "This was our first sign that Celera was setting out to undermine the international effort. What is it that motivates Craig? I think he's motivated by the same things that drive other scientists — personal ego, a degree of altruism, a desire to push human knowledge forward — but there must be something else that drives the guy. I think Craig has a huge chip on his shoulder that makes him want to be loved. I actually think Craig is desperate to win a Nobel Prize. He also wants to be very, very rich. There is a fundamental incompatibility there."

One day, I ran into a young player in the Human Genome Project. He believed in the worth and importance of the project, and said that he had turned down a job offer from Celera. He didn't have any illusions about human nature. He said, "Here's why everyone is so pissed at Craig. The whole project started when James Watson persuaded Congress to give him money for the human genome, and he turned around and gave it to his friends — they're the heads of centers today. It grew into a lot of money, and then the question was, Who was going to get the Nobel Prize? In the United States, there were seventeen centers in the project, and there was no quality control. It didn't matter how bad your data was, you just had to produce it, and people weren't being held accountable for the quality of their product. Then Celera appeared. Because of Celera, the NIH was suddenly forced to consolidate its funding. The NIH and Francis Collins began to dump more than 80 percent of the money into just three centers — Baylor, Washington University, and MIT — and they jacked everybody else. They had to do it, because they had to race Celera, and they couldn't control too many players. So all but three centers were cut drastically, and some of the labs closed down. Celera was not just threatening their funding but threatening their very lives and everything they had spent years building. It's kind of sad. Now those people hang around meetings, and the leaders treat them like 'If you're really nice, we'll give you a little piece of the mouse.' That's the reason so many of them are so angry at Celera. It's easier for them to go after Craig than to go after Francis Collins and the NIH."

At Celera's headquarters in Rockville, I was shown how human DNA was shotgunned into small pieces when it was sprayed through a hospital nebulizer that cost $1.50. The DNA fragments were then introduced into *E. coli* bacteria, and grown in glass dishes. The bacteria formed brown spots — clones — on the dishes. Each spot had a different fragment of human DNA growing in it. The dishes were carried to a room where three robots sat in glass chambers the size of small bedrooms. Each robot had an arm that moved back and forth rapidly over a dish. Little needles on the arms kept stabbing down and taking up the brown spots.

Craig Venter stood watching the robots move. The room smelled faintly like the contents of a human intestine. "This used to be done by hand. We've been picking 55,000 clones a day," he said. (Later, Celera got that rate up to 120,000 clones a day.) All the DNA fragments would eventually wind up in the Prism sequencing machines, and what would be left, at the end, was a collection of up to 22 million random fragments of sequenced human DNA. Then the river of shattered DNA would come to the computer, and to a computer scientist named Eugene Myers, who with his team devised the First Assembly.

Gene Myers has dark hair and a chiseled, handsome face. He wears glasses and a green half-carat emerald in his left ear and brown Doc Martens shoes. He also has a ruby and a sapphire that he will wear in his ear, instead of the emerald, depending on his mood. He is sensitive to cold. On the hottest days of summer, Myers wears a yellow Patagonia fiberpile jacket, and he keeps a scarf wrapped around his neck. "My blood's thin," he explained to me. He says the scarf is a reference to the DNA of whatever organism he happens to be working on. When I first met Myers, in the hot summer of 1999, he was keeping himself warm in his fruit-fly scarf. It had a black-and-white zigzag pattern. This spring, Myers started wearing his human scarf, which has a green chenille weave of changing stripes. He intended his scarf to make a statement about the warfare between Celera and the public project. "I picked green for my human scarf because I've heard that green is a positive, healing color," he said. "I really want all this bickering to go away." His office is a cubicle in a sea of cubicles, most of which are stocked with Nerf guns, Stomp Rockets, and plastic Viking helmets. Occasionally, Myers puts the "Ride of the Valkyries" on a boom box, and in a loud voice he declares war. Nerf battles sweep through Celera whenever the tension rises. Myers fields a compound double-action Nerf Lock-N-Load Blaster equipped with a Hyper*Sight. "Last week we slaughtered the chromosome team," he said to me.

Myers used to be a professor of computer science at the University of Arizona in Tucson. He specializes in combinatorial algorithms. This involves the arrangements and patterns of objects. One day in 1995, he got a telephone call from a geneticist named James Weber, at the Marshfield Medical Research Foundation in Wisconsin. Weber said he felt that whole-genome shotgunning would work for organisms that have very long DNA molecules, such as humans. He wondered it Myers could help him with the math.

Jim Weber submitted a proposal to the NIH for a grant — $12 million — to support a pilot study of the shotgun approach on the human genome. This might speed up the project dramatically, he suggested. Weber was invited to speak to the annual meeting of the heads of the project, held in Bermuda.

Weber was nervous about it, and wanted Gene Myers to go with him to help explain the math. "Jim asked them to invite me, but they didn't," Myers says. So on February 26, 1996, Jim Weber went alone to the meeting in Bermuda and tried to make a case for shotgunning the human genome. He found himself facing a U-shaped table with about forty people at it. "They trounced Jim," Myers said. "They said it wouldn't work. They said it would be full of holes. 'A Swiss-cheese genome' — that's the term we've often heard. The grant proposal was soundly rejected."

Jim Weber says that the Swiss-cheese analogy was not far off, but that "it would have been much better to get most of the human genome quickly, even with holes in it, so that people could start using the information to understand diseases and begin to find cures for them. It would have been better if the NIH had funded a pilot study. Instead, Gene and his team went out and did it. That is a huge accomplishment."

Craig Venter was hanging around while I was talking with Myers. He came up to us and said, "They not only shot Gene down — they ridiculed him. They said he was a kook. We're going to prove that Gene was right, and we're going to prove that there's something fundamentally wrong with the system."

On September 9, 1999, Venter announced that Celera had completed the sequencing of the fruit fly's DNA, and had begun to run human DNA through its sequencing machines — there were now

300 of them crammed into Building One in Rockville. The Command Center was up and running, and from then on Celera operated in high-speed mode. One day that fall, I talked with the company's information expert, a stocky man named Marshall Peterson. He took me to the computer room, in Building Two. To get into the room, Peterson punched in a security code and then placed his hand on a sensor, which read the unique pattern of his palm. There was a clack of bolts sliding back, and we pushed through the door.

A chill of cold air washed over us, and we entered a room filled with racks of computers that were wired together. "What you're looking at in this room is roughly the equivalent of America Online's network of servers," Peterson said. "We have fifty-five miles of fiber-optic cables running through this building." Workmen standing on ladders were installing many more cables in the ceiling. "The disk storage in this room is five times the size of the Library of Congress. We're getting more storage all the time. We need it."

He took me to the Command Center, where a couple of people were hanging around consoles. Some of the consoles had not had equipment installed in them yet. A big screen on the wall showed CNN *Headline News.* "I've got a full-time hacker working for me to prevent security breaches," Peterson said. "We're getting feelers over the Internet all the time — people trying to break into our system." Celera would be dealing with potentially valuable information about the genes of all kinds of organisms. Peterson thought that some of what he called feelers — subtle hacks and unfriendly probes — had been emanating from Celera's competitors. He said he could never prove it, though. Lately, the probes had been coming from computers in Japan. He thought it was American hackers coopting the Japanese machines over the Internet.

By October 20, forty days after Celera started running human DNA through its machines, the company announced that it had sequenced 1.2 billion letters of human code. The letters came in small chunks from all over the genome. Six days later, Venter announced that Celera had filed provisional patent applications for 6,500 human genes. The applications were for placeholder patents. The company hoped to figure out later which of the genes would be worth patenting in earnest.

A gene patent gives its holder the right to make commercial

products and drugs derived from the gene for a period of seventeen years. Pharmaceutical companies argue that patents are necessary, because without them businesses would never invest the hundreds of millions of dollars that are needed to develop a new drug and get it through the licensing process of the Food and Drug Administration. ("If you have a disease, you'd better hope someone patents the gene for it," Venter said to me.) On the other hand, parceling out genes to various private companies could lead to what Francis Collins refers to as the "Balkanization of the human genome," a paralyzing situation that might limit researchers' access to genes.

Venter insists that Celera is an information company and that patenting genes is not its main goal. He has said that Celera will attempt to get patents on not more than about 300 human genes. There is no question that Celera hopes to nail down some very valuable genes — billion-dollar genes, perhaps.

Celera's stock had drifted since the summer, but around Halloween, as investors began to realize that the company was cranking out the human genome — and filing large numbers of placeholder patents — it jumped up to forty dollars a share. (The prices here are pre-split prices. Adjusted for today's prices, the stock moved up to twenty dollars.) On December 2, the Human Genome Project announced that it had deciphered most of the code on chromosome No. 22, the second-shortest chromosome in the human genome. This made the reading of the whole genome seem more imminent, and Celera's stock began a spectacular rise. It shot up that day by nine points, to close at over seventy dollars. Then, after the market's close on Thursday, December 16, Jeff Fischer, a cofounder of the Web site called The Motley Fool, announced that he was buying shares of Celera for his own portfolio. It is called the Rule Breaker Portfolio, and it has famously delivered wealth — Fischer bought AOL very early, for example. On that Friday morning, a great number of people tried to buy Celera, and they drove the stock up twenty points. It was on its way to the pre-split equivalent of more than $500 a share. That past summer, it had been trading at fourteen.

I went to visit Celera on Tuesday of the following week, and that morning the company's stock could not open for trading. Everyone wanted to buy it, and nobody wanted to sell it. While the stock was halted — at $101 a share — I wandered around. There was a feeling of paralysis in the air, and I sensed that not much work was getting done that day, except by the machines. Employees were checking the quote on the Internet and wondering what their net worth would be when the stock opened. The lobby now sported fisheye security cameras. The walls smelled of fresh paint, and the floors had a new purple carpet with a pattern that resembled worms. They were meant to look like fragments of DNA.

I found Hamilton O. Smith in his lab, puttering around with human DNA in tiny test tubes, but his heart was not in the job. He was tired. He explained that he was renovating an old house that he and his wife had bought. He had stayed up all night ripping carpet out of the basement, because new carpets were due to arrive that morning. He had driven to work in his '83 Mercury Marquis. He owned thousands of shares of Celera.

Smith passed a computer, stopped, and brought up a quote. Celera had finally opened for trading. It had gapped up — jumped instantly upward — by thirteen points. It was at 114. Smith's net worth had gapped up by something on toward a million dollars. "Is there no end to this?" he muttered.

Craig Venter came into Smith's lab and asked him to lunch. In the elevator, Smith said to him, "I can't stand it, Craig. The bubble will break." They sat down beside each other in the cafeteria and ate cassoulet from bowls on trays.

"This defies common sense," Smith said. "It's really impossible to put a value on this company."

"That's what we've been telling the analysts," Venter said.

Later that day, I ended up in Claire Fraser's office at TIGR headquarters, a complex of semi-Mission-style buildings a couple of miles from Celera's offices and labs. Fraser is a tall, reserved woman with dark hair and brown eyes, and her voice has a faint New England accent. She grew up in Saugus, Massachusetts. In high school, she says, she was considered a science geek. "The only lower citizens were the nerdy guys in the audiovisual club. Of course, now they're probably in Hollywood." Her office has an Oriental rug on the floor and a table surrounded by Chippendale chairs. It was originally Venter's office. ("This is Craig's extravagant taste, not mine," she explained.) She wore an expensive-looking suit. Two poodles, Cricket and Marley, slept by a fireplace.

"Before genomics, every living organism was a black box," she said. "When you sequence a genome, it's like walking into a dark room and turning on a light. You see entirely new things everywhere."

Fraser placed a sheet of paper on the table. It contained an impossibly complicated diagram that looked like a design for an oil refinery. She explained that it was an analysis of the genome of cholera, a single-celled microbe that causes murderous diarrhea. TIGR scientists had finished sequencing the organism's DNA a few weeks earlier. Much of the picture, she said, was absolutely new to our knowledge of cholera. About a quarter of the genes of every microbe that had been decoded by TIGR were completely new to science, and were not obviously related to any other gene in any other microbe. To the intense surprise and wonder of the scientists, nature was turning out to be an uncharted sea of unknown genes. The code of life was far richer than anyone had imagined.

Fraser's eyes moved quickly over the diagram. "Yes . . . wow. . . . There may be important transporters here. . . . You see these transporters in other bacteria, and . . . I don't know . . . it looks like there could be potential for designing a new drug that could block them.

The phone rang. Fraser walked across her office, picked up the receiver, and said softly, "Craig? Hello. What? It closed at a hundred and twenty-five?" Pause. "I don't know how much it's worth — you're the one with the calculator."

Their net worth had jumped above $150 million that day.

Fraser drove home, and I followed her in my car. Their house is in the country outside Washington. It sits behind a security gate at the end of a long driveway. Venter arrived in a new Porsche. The car would do zero to sixty in five seconds, he said. In the vaulted front hall of the house there was a large stained-glass window showing branches of a willow tree, and there was a model of HMS *Victory* in a glass case. A jumble of woodworking machines — a band saw, a table saw, a drill press — filled a shop attached to the garage. Venter has worked with wood since high school.

In the kitchen, Claire fixed dinner for the poodles, while Craig circled the room, talking. "We created close to two hundred millionaires in the company today; I think most of them had not a clue this would happen when they joined Celera. We have a secretary who became a millionaire today. She's married to a retired policeman. He went out looking to buy a farm." He popped a Bud Light and swigged it. "This could only happen in America. You've got to love this country." Claire fed the poodles.

There were no cooking tools in the kitchen that I could see. The counters were empty. The only food I noticed was a giant sack of dog food, sitting on top of an island counter, and two boxes of cold cereal — Quaker Oatmeal Squares and Total. In the guest bathroom, upstairs, there were no towels, and the walls were empty. The only decorative object in the bathroom was a cheap wicker basket piled with little soaps and shampoos they had picked up in hotels.

We went to a restaurant and ate steak. "We're in the Wild West of genomics," Venter said. "Celera is more than a scientific experiment; it's a business experiment. Our stock-market capitalization as of today is three and a half billion dollars. That's more than the projected cost of the Human Genome Project. I guess that's saying something. The combined market value of the Big Three genomics companies — Celera, Human Genome Sciences, and Incyte — was about twelve billion dollars at the end of today. This wasn't imaginable six months ago. The Old Guard doesn't have control of genomics anymore." He chewed steak, and looked at his wife. "What the hell are we going to do with all this money? I could play around with boats. . . ."

Claire started laughing. "My God, I couldn't live with you."

"The money's nice, but it's not the motivation," Venter said to me. "The motivation is sheer curiosity."

In December 1999, Celera and the Human Genome Project discussed whether it would be possible to collaborate. There was one formal meeting, and there were many points of difference. Meanwhile, Celera's stock seemed to go into escape velocity from the earth. In January, it soared over $200 a share. Celera filed to offer more shares to the public and declared a two-for-one stock split. Shortly after the split, on February 25, the stock hit an all-time high of $276 a share (more than $550, pre-split). Celera's stock-market value reached $14 billion, and Venter's worth surpassed $700 million. It looked as if Venter could become a billionaire of biotechnology.

Then, on March 6, newspapers carried reports that the discussions between Celera and the public project had collapsed. The main point of disagreement, according to officials at the public project, was that Celera wanted to keep control of intellectual property in the human genome. Celera intended to license its analyzed database to pharmaceutical companies and nonprofit research institutions, for payment. Celera said that it would let anyone use the data, but that any other company would be forbidden from reselling the data. The Human Genome people insisted that the period of restriction on the data could be for no more than a year, and after that the data should be totally public. Celera argued that it didn't want its competitors to resell the information and profit from Celera's work. Celera's stock began to drop.

On March 14, President Clinton and Prime Minister Blair of Great Britain released a joint statement to the effect that all the genes in the human body "should be made freely available to scientists everywhere." The statement had been drafted with the help of Francis Collins and his staff, and had been in the works for a year. It was vague, but it looked like an Anglo-American smart bomb aimed at Celera, and it scared the daylights out of investors in biotechnology stocks, who feared that potentially lucrative patents on genes might be undermined by some new government policy.

On the day of the Clinton-Blair statement, Celera's stock went into a screaming nosedive. It dropped $57 in a matter of hours, amid trading halts and order imbalances. The other genomic stocks crashed in sympathy with Celera, and this, in turn, dragged down the Nasdaq, which that day suffered the second-largest point loss in its history. Short-sellers — people who profit from the decline of a stock — encrusted Celera like locusts. As of this writing, the Nasdaq has not recovered. Venter's mother telephoned him afterward, and said to him, "Craig, you've managed to do overnight what Alan Greenspan has been trying to do for years."

"It's not every day you get attacked by the president and the prime minister," Venter said to me late that night on the telephone. "I'm expecting a call from the pope any day now, asking me to recant the human genome." He sounded wired and exhausted. "I feel a little like Galileo. They offered to have a barbecue with him, right? Look, I'm not likening myself to Galileo in terms of genius, but it is clear that the human genome is the science event of our time. I am going to publish the genome, and that's what the threat to the public order is. If Celera was keeping the genome a secret, the way Incyte and Human Genome Sciences are, you wouldn't hear a peep out of the government. Our publishing the genome makes a mockery of the fifteen years and billions of dollars the public project has spent on it."

Venter seemed particularly upset with the British part of the public project. "In my opinion," he said, "the Wellcome Trust is now trying to justify how, as a private charity, it gave what I think was well over a billion dollars to the Sanger Centre to do just a third of the human genome, largely at the expense of the rest of British medical science. Clinton and Blair took forty billion dollars out of the biotechnology industry today — that's how much was lost by investors. It was money that would pay for cures for cancer, and it was taken off the table, all because some bastards at the Wellcome Trust are trying to cover up their losses."

I called Michael Morgan, at the Wellcome Trust, to see what he had to say about this. "In hindsight, it is easy to ascribe to us Machiavellian powers that the Prince would have been proud of," he said dryly. "As for the allegation that I'm a bastard, I can easily disprove it using the technology of the Human Genome Project."

The day after the Clinton-Blair statement and the crash in biotech stocks, a White House spokesman made a point of telling reporters that the administration supported the patenting of genes.

On March 24, Venter and his colleagues published a substantially complete genome of the fruit fly — *Drosophila* — in *Science*. It was also published by Celera on a CD, which Venter had placed on the chairs of 1,300 fly researchers at a conference in Pittsburgh. Venter emphasized the fact that the fly genome had been a collaboration with a publicly funded project. In other words, he was suggesting there was no real reason that the Human Genome Project couldn't collaborate with Celera, too. The fly project — known as the Berkeley Drosophila Genome Project — is headed by a fly geneticist named Gerald Rubin.

"One of the things I really like about Craig Venter is that he almost totally lacks tact," Rubin said to me. "If he thinks you are an idiot, he will say so. I find that way of dealing very enjoyable. Craig is like somebody who's using the wrong fork at a fancy dinner. He'll tell you what he thinks of the food, but he won't even think about what fork he's using. It was a great collaboration."

John Sulston, the head of the Sanger Centre, told the BBC that he felt Celera planned to "Hoover up all the public data, which we are producing, add some of their own, and sell it as a packaged product." He added, "The emerging truth is absolutely extraordinary. They really do intend to establish a complete monopoly position on the human genome for a period of at least five years," and he said, "It's something of a con job."

"Sulston essentially called us a fraud. It's like he's been bit by a rabid animal," Venter fumed.

"It's puzzling. To me, the whole fight defies rational analysis," Hamilton O. Smith said to me, shortly after his net worth had cratered in Celera's mudslide. "But the publicly funded labs are angry for reasons I can partly understand. We took it away from them. We took the big prize away from them, when they thought they would be the team that would do the whole human genome and go down in history. Pure and simple, they hate us."

On April 6, Venter announced that Celera had finished the sequencing phase of the human genome, and was moving on to First Assembly. Celera had produced some 18 million fragments of the first genome, perhaps Craig Venter's. Soon afterward, on an unseasonably warm day, while the cherry trees were in full blossom, I visited Celera to see how the assembly was going. I found Gene Myers in his cubicle, looking chilled. He was bundled up in his yellow fiberpile jacket and his green human scarf. He and his team had started running chunks of human DNA code through the computers over the weekend. The first run had resulted in a mess — something was wrong with the software. They had done some tweaks, and they were running a few more chunks of code. It would take months to assemble the whole thing.

"Assembly is pretty boring," he said, somewhat apologetically.

Myers said that Celera would be using all of the Human Genome Project's human DNA code — which was published on the GenBank Web site — and would tear it into fragments and compare them with Celera's DNA code, and then the software that his team had written would try to assemble all the fragments into a whole human genome. At the same time, Celera was coolly telling the public project that its scientists could see Celera's data but only if they came to look at the data on Celera's computer. The collaboration had never come to pass.

Minutes later, one of Myers's people, a computer scientist named Knut Reinert, hurried in, and told him that the first assembled human genome sequence had just come out of the computers. Myers put the "Ride of the Valkyries" on the boom box, and fifteen people tried to crowd into Reinert's cubicle.

Myers bent over Reinert's shoulder and said, "We got it! We got the first one! This is the first assembled human sequence we've gotten out of nature!"

What appeared on the screen was a mathematical diagram of a stretch of human DNA. It showed arrows going in various directions, connecting dots together. "The picture looks like a Super Bowl debriefing," one Celera programmer remarked.

They talked about it for a few minutes, and then everyone drifted back to work. That day, Celera's stock dropped another 20 percent.

In early May, another company got into the business of the human genome. DoubleTwist, Inc., announced that it had teamed up with Sun Microsystems to compete with Celera. DoubleTwist and Sun were offering an analyzed database of the human genome to anyone for a fee, using the data from the Human Genome Project, not from Celera. The price was $650,000 for a database that would be updated regularly.

At the same time, Celera's stock had gone down below $100 a share. Many investors had recently bought the secondary offering, paying $225 a share for it. This brought on a slew of class-action lawsuits against Celera, filed by law firms specializing in shareholder suits. There were various claims, sparked by the fact that the secondary investors had lost 60 percent of their money. These lawsuits will probably be consolidated, and Celera will either settle or fight them in court.

As for the science, knowledgeable observers believed that, in the end, Celera had actually spent about half a billion dollars to sequence the human genome — 300 million more than Venter had originally predicted. Was it worth it? I asked many biologists about this, and most of them spoke the way scientists do when they believe that a great door has been opened, and light is shining deep into nature, suggesting the presence of rooms upon rooms that have never been seen before. There was also a clear sense that the door would not have been opened so soon if Craig Venter and Celera had not given it a swift kick.

"We can thank Venter in retrospect," James Watson said, leaning back and smiling and squinting at the ceiling. "I was worried he could do it, and that would stop public funding of the Human Genome Project. But if an earthquake suddenly rattled through Rockville and destroyed Celera's computers, it wouldn't make much difference." He stood up, and offered me the door.

Eric Lander, who professes to like Venter, said, "Having the human genome is like having a Landsat map of the earth, compared to a world where the map tapers off into the unknown, and says, 'There be dragons.' It's as different a view of human biology as a map of the earth in the 1400s was compared to a view from space today." As for the war between Celera and the public project, he said, "At a certain level, it is just boys behaving badly. It happens to be the most important project in science of our time, and it has all the character of a schoolyard brawl."

Norton Zinder, Watson's friend, who had feared that he would die before he saw the human genome, said that he felt marvelous. "I made it. Now I've gotta stay alive for four more years, or I won't get all my options in Celera." Zinder, who is a vigorous-seeming older man, was sprawled in a chair in his office overlooking the East River, gesturing with both hands raised. He shifted gears and began to look into the future. "This is the beginning of the beginning," he said. "The human genome alone doesn't tell you crap. This is like Vesalius. Vesalius did the first human anatomy." Vesalius published his work in 1543, an anatomy based on his dissections of cadavers. "Before Vesalius," Zinder went on, "people didn't even know they had hearts and lungs. With the human genome, we finally know what's there, but we still have to figure out how it all works. Having the human genome is like having a copy of the Talmud but not knowing how to read Aramaic."

DAVID QUAMMEN

Megatransect

FROM *National Geographic*

AT 11:22 ON THE MORNING of September 20, 1999, J. Michael Fay strode away from a small outpost and into the forest in a remote northern zone of the Republic of the Congo, setting off on a long and peculiarly ambitious hike. By his side was an aging Pygmy named Ndokanda, a companion to Fay from adventures past, armed now with a new machete and dubiously blessed with the honor of cutting trail. Nine other Pygmies marched after them, carrying waterproof bags of gear and food. Interspersed among that troop came still other folk — a camp boss and cook, various assistants, Michael "Nick" Nichols with his cameras, and me.

It was a hectic departure to what would eventually, weeks and months later, seem a quiet, solitary journey. Fay planned to walk across central Africa, more than a thousand miles, on a carefully chosen route through untamed regions of tropical forest and swamp, from northeastern Congo to the coast of Gabon. It would take him at least a year. He would receive resupply drops along the way, communicate as needed by satellite phone, and rest when necessary, but his plan was to stay *out there* the whole time, covering the full route in a single uninterrupted push. He would cross a northern stretch of the Congo River Basin, then top a divide and descend another major drainage, the Ogooué.

Any big enterprise needs a name, and Fay had chosen to call his the Megatransect — *transect* as in cutting a line, *mega* as in mega, a label that variously struck those in the know as amusing or (because survey transects in field biology are generally straight and involve statistically rigorous repetition) inappropriate. Fay is no sobersides, but amusement was not his intent. Behind this mad lark lay a serious purpose — to observe, to count, to measure, and from those observations and numbers to construct a portrait of great central African forests before their greatness succumbs to the inexorable nibble of humanity. The measuring began now. One of Fay's entourage, a bright young Congolese named Yves Constant Madzou, paused at the trailhead to tie the loose end of a string to a small tree.

I paused beside him, intrigued. In the technical lingo this string was known as a topofil. Its other end was wound on a conical spool inside a Fieldranger 6500, a device used by foresters for measuring distance along any walked route. The topofil pays out behind a walker while the machine counts traversed footage, much as a car's odometer counts miles. Each spool holds a 6-kilometer (3.7-mile) length. Madzou carried a half dozen extras, and somewhere among the expedition supplies were many more. Having served its purpose each day, the string would quickly disappear down the gullets of termites and other jungle digesters, I'd been told, but the numbers it delivered with such Hansel and Gretel simplicity would be accurate to the nearest twelve inches. You can't get that precision from a global positioning system (GPS) and a map. Running the topofil was to be one of Madzou's assignments.

Now, as he stepped out after Fay in the first minutes of Day 1, the Fieldranger gurgled in a low, wheezy tone. Madzou trailed filament like a spider. The string hovered, chest high, under tension. And I found it piquant to contemplate that if Fay's expedition proceeds to its fulfillment, a thousand-mile length of string will go furling out through the equatorial jungle.

That string seemed an emblem of all the oxymoronic combinations this enterprise embodies — high tech and low tech, vast scales and tiny ones, hardheaded calculation and loony daring, strength and fragility, glorious tropical wilderness and a mitigated smidgen of litter. As he walks, Fay will gather data in many dimensions by many means, including digital video camera, digital audio recorder, digital still camera, GPS, thermohygrometer (whatever that is), handheld computer, digital calipers, hand lens (for squinting at tiny specimens), and notebook and pencil. The topofil will be a quaint but important complement to the rest.

Within less than an hour on the first day we're shin-deep in mud, crossing the mucky perimeter of a creek. "Doesn't take long for the swamps to kick in around here," Fay says cheerily. He's wearing his usual outfit for a jungle hike: river sandals, river shorts, a lightweight synthetic T-shirt that can be rinsed out each evening and worn again next day, and the day after, and every day after that until it disintegrates. River sandals are preferable to running shoes or tall rubber boots, he has found, because the forest terrain of northeastern Congo is flat and sumpy, its patches of solid ground interlaced with leaf-clotted spring seeps and black-water creeks, each of them guarded by a corona of swamp. A determined traveler on a compass-line march is often obliged to wallow through sucking gumbo, cross a waist-deep channel of whiskey-dark water flowing gently over a bottom of white sand, wallow out through the muck zone on the far side, rinse off, and keep walking. Less determined travelers, in their Wellingtons and bush pants, just don't get to the places where Fay goes.

He stops to enter a datum into his yellow "Rite in the Rain" notebook: elephant dung, fresh. Blue and black swallowtail butterflies flash in sun shafts that penetrate the canopy. He notes some fallen fruits of the plant *Vitex grandifolia.* Trained as a botanist before he shifted focus to do his doctorate on western lowland gorillas, Fay's command of the botanical diversity upon which big mammals depend is impressive — he seems familiar with every tree, vine, and herb. He knows the feeding habits of the forest elephant (*Loxodonta africana cyclotis,* the smaller subspecies of African elephant adapted to the woods and soggy clearings of the Congo Basin) and the life cycles of the plants that produce the fruits it prefers. He can recognize, from stringy fecal evidence, when a chimpanzee has been eating a certain latex-rich fruit. He can identify an ambiguous tree by the smell of its inner bark. He sees the forest in its particulars and its connectedness. Now he bends pensively over a glob of civet dung. Then he makes another notation.

"Mmm. This is gonna be fun," he says, and walks on.

Mike Fay isn't the first half-crazed white man to set out trekking across the Congo Basin. In a tradition that includes such Victorian-era explorers as David Livingstone, Verney Lovett Cameron, Savorgnan de Brazza, and Henry Morton Stanley, he's merely the latest. Like Stanley and some of the others, he has a certain gift for command, a level of personal force and psychological savvy that allows him to push a squad of men forward through difficult circumstances, using a mix of inspirational goading, promised payment, sarcasm, imperiousness, threat, tactical sulking, and strong example.

He's a paradoxical fellow and therefore hard to ignore, a postmodern redneck who chews Red Man tobacco, disdains political correctness, knows a bit about tractor repair and a lot about software, and views the suburbanized landscape of modern America with cold loathing. Born in New Jersey, raised there and in Pasadena, California, he has fully transplanted his sense of what's home. "I plan on dying out here," he says of the central African forest. "I'll never go back to live in the U.S."

What makes Fay different from those legendary Victorian zealots is that he's not traveling in service of God or empire or for the personal enrichment of the king of Belgium. He does have backers, most notably the Wildlife Conservation Society (WCS) of New York, for which he's a staff member on field assignment, and the National Geographic Society, but he's certainly not laboring for the greater glory of them. His driving motive — or rather the first and more public of his two driving motives — is conservation.

His immediate goal is to collect a huge body of diverse but intermeshed information about the biological richness of the ecosystems he'll walk through and about the degree of human presence and human impact. He'll gather field notes on the abundance of elephant dung, leopard tracks, chimpanzee nests, and magisterial old-growth trees. He'll make recordings of birdsong for later identification by experts. He'll store away precise longitude-latitude readings (automatically, every twenty seconds throughout the walking day) with his Garmin GPS unit and the antenna duct-taped into his hat. He'll detect gorillas by smell and by the stems of freshly chewed *Haumania danckelmaniana*, a tangly monocot plant they munch like celery. Eventually he will systematize those data into an informational resource unlike any ever before assembled on such a scale — with the ultimate goal of seeing that resource used wisely by the managers and the politicians who will decide the fate of African landscapes.

"It's not a scientific endeavor, this project," Fay acknowledges during one of our talks before departure. Nor is it a publicity stunt, he argues, answering an accusation that's been raised. What he means to do, he explains, is to "quantify a stroll through the woods."

Then there's his second driving motive. He doesn't voice it explicitly, but I will: Mike Fay is an untamable man who just loves to walk in the wilds.

Completing this thousand-mile trek won't be easy, not even for him. There are dire diseases, armed poachers, political disruptions, and other sorts of threat and mishap that could stop him. He's familiar with malaria and filariasis, aware of Ebola, and has found himself inconveniently susceptible to foot worms, a form of parasite that can travel from elephant dung into exposed human feet, burrowing tunnels in a person's toes, only to die there and fester. But the biggest challenge for Fay will come *after* all his walking.

Can he make good on the claim that this encyclopedia of field data will be useful? Can he channel his personal odyssey into practical results for the conservation of African forests?

Suddenly, a mile and a quarter on, Fay makes a vehement hand signal: *stop.* As we stand immobile and hushed, a young male elephant appears, walking straight toward us through the understory. Ndokanda slides prudently to the back of the file, knowing well that a forest elephant, nearsighted and excitable, is far more dangerous than, say, a leopard. Fay raises the video camera. The elephant, visually oblivious and upwind of our smell, keeps coming. The videotape rolls quietly. When the animal is just fifteen feet from him and barely twice that from the rest of us — too close for anyone's comfort — Fay says in a calm voice: "Hello." The elephant spooks, whirls around, disappears with its tail streaming high.

Tusk length, about forty centimeters, Fay says. Maybe ten or twelve years old, he estimates. It goes into his notebook.

Fay is a compact forty-three-year-old American with a sharp chin and a lean, wobbly nose. Behind his wire-rimmed glasses, with their round, smoky lenses, he bears a disquieting resemblance to the young Roman Polanski. Say something that's doltish or disagreeable, and he'll gaze at you silently the way a heron, hungry or not, gazes at a fish. But on the trail he's good company, a man of humor and generous intellect. He sets a punishing pace, starting at daylight, never stopping for lunch or rest, but when there are field data to record in his yellow notebook, fortunately, he pauses often.

We leave camp just after dawn on Day 3 and follow the Mopo River downstream along a network of elephant trails. We're a smaller group now, Nick Nichols and his assistant having gone off to photograph some Bambendjellé Pygmy groups, intending to rendezvous with Fay several weeks later. Fay, Madzou, and I set out while the crew are still eating breakfast, giving us a relatively quiet first look at forest activity. Under a high canopy of *Gilbertiodendron dewevrei* trees, the walking is easy. The understory is sparse and well trampled by elephant traffic, as it generally tends to be in these dominant stands of *G. dewevrei,* more familiar to the Pygmies as *bemba.* Later, as we swing away from the river onto higher ground, the bemba gives way to a mixed forest, its canopy gaps delivering light to a clamorous undergrowth of brush, saplings, vines, thorny monocots, and woody lianas, through which we climb hunchbacked behind the day's point man. The thickest zones of such early successional vegetation are known in the local language as *kaka,* a coincidental but appropriate convergence with the scatological slang term from various other languages. Kaka to a Pygmy, "crappy forest" to you and me. Today it's Bakembe, younger and stronger than Ndokanda, who cuts us a tunnel through the kaka.

The most devilish of the thorny plants is *Haumania danckelmaniana,* mentioned already as a favored gorilla food. Looping high and low throughout the understory, weaving the kaka into a tropical brier patch, forever finding chances to carve bloody scratches across unprotected ankles and toes, *H. danckelmaniana* is the bushwhacker's torment. Even a Congo walker as seasoned as Fay has to spend much of his time looking down, stepping carefully, minimizing the toll on his feet. Of course Fay would be looking down anyway, because that's where so much of the data are found — scat piles, footprints, territorial scrape marks, masticated stems, serpentine tracks left by red river hogs nose-plowing through leaf litter, pangolin burrows, aardvark burrows, fallen leaves, fallen fruit. Fay's GPS tells us where we are, while his map and our compasses tell us which way to go. We see no human trails in this forest, because there are no resident humans, few visitors, and no destinations.

Fay pauses over a pile of gorilla dung, recognizing seeds of *Maranthes glabra* as a hint about this animal's recent diet. Farther on he notes the hole where a salt-hungry elephant has dug for minerals. Farther still the print of a yellow-backed duiker, one of the larger forest antelope. Each datum goes into the notebook, referenced to the minute of the day, which will be referenced in turn by his GPS to longitude and latitude at three decimal points of precision. Years from now, his intricate database will be capable of placing that very pile of gorilla dung at its exact dot in space-time, should anyone want to know.

When it comes time to ford the Mopo, Fay wades knee-deep into the channel with his video camera pressed to his face. Spotting a dark lump against the white sand, he gropes for it one-handed, still shooting. "*Voilà.* A palm nut." He shows me the hard, striated lump, smaller than a walnut, light in weight but heavy with import. It's probably quite old, he explains. He has found thousands like this in his years of wading the local rivers, and carbon-dating analysis of a sizable sample revealed them to be durable little subfossils, ranging in age between about 1,000 and 2,400 years. Presumably they wash into streams after centuries of shallow burial in the soil nearby. What makes their presence mysterious is that this species of palm, *Elaeis guineensis,* is known mainly as an agricultural species, grown on plantations near traditional Bantu villages at the fringe of the forest and harvested for its oil. *Elaeis guineensis* seems to need cleared land, or at least edges, and to be incapable of competing in dense, mature forest.

The abundance of ancient oil palm nuts in the river channels suggests a striking possibility: that a vast population of proto-Bantu agriculturalists once occupied this now vacant and forested region. So goes Fay's line of deduction anyway. He hypothesizes that they cut the forest, established palm plantations, discarded millions or billions of palm nuts in the process of extracting oil, and then vanished mysteriously. Some scholars argue that natural climate change over the past three millennia might account for the coming and going of oil palms, the natural ebb and return of forest, but to Fay it doesn't make sense, given the apparent dependence of *Elaeis guineensis* on human agriculturalists. "What makes sense," he says, "is that people moved in here, grew palm nuts, and then died out." Died out? From *what?* He can only guess: maybe warfare, or a killer drought, or population overshoot leading to ecological collapse. Or maybe an early version of AIDS or Ebola emptied the region of people, more or less abruptly, allowing the forest to regrow.

There's no direct evidence for this cataclysmic depopulation, but it's a theme that will recur throughout Fay's hike. Meanwhile, he drops the palm nut into a Ziploc bag.

Beyond the Mopo we sneak up on a group of gorillas feeding placidly in a *bai,* a boggy clearing amid the forest. We approach within a hundred feet of an oblivious female as she works her way through a salad of *Hydrocharis* stems, nipping off the tender white bases, tossing the rest aside. Her face is long and tranquil, with dark eyes shaded beneath the protrusive brow. The hair on her head is red, Irish red, as it often is among adult lowland gorillas. Her arms are huge, her hands big and careful. Leaving me behind, Fay skulks closer along the bai's perimeter. When the female raises her head to look straight in his direction, the intensity of her stare seems to bring the whole forest to silence. For a minute or two she looks puzzled, wary, menacingly stern. Then she resumes eating. Fay gets the moment on zoom-lens video. Later he tells me that he froze every muscle during those seconds she glowered al him, not daring to lower the camera, not daring to move, while a tsetse fly sucked blood from his foot.

The video camera, with its soundtrack for verbal annotations and its date-and-time log, is becoming one of his favorite tools. He shoots footage of major trees, posing a Pygmy among the buttresses for scale. He shoots footage of monitor lizards and big unidentified spiders. He shoots footage, for the hell of it, of me floundering waist-deep in mud. Occasionally he does a slow 360-degree pan to show the wraparound texture of a patch of forest. And when I alert him that a leech has attached itself to one of the sores on his right ankle, he videos that. Then he hands me the camera, while Madzou burns the leech off with a lighter, so I can capture the operation from a better angle.

At noon he inspects another fresh mound of elephant dung, poking his finger through the mulchy gobs. Elephants here eat a lot of fallen fruit, but just what's on the menu lately? He picks out seeds of various shape and size, identifying each at a glance, reciting the Latin binomials as he tosses them into a pile: "*Panda oleosa, Tridesmostemon omphalocarpoides, Antrocaryon klaineanum, Duboscia macrocarpa, Tetrapleura tetraptera, Drypetes gossweileri,* and what's this other little thing, can't remember, wait, wait . . . oh yeah, *Treculia africana.*"

As I squat beside him, impressed by his knowledge and scribbling the names, he adds: "Of course, this is where you get foot worms, standing in elephant dung like this."

We make camp along a tributary of the Mopo. The Pygmies erect a roof beam for the main tarp and a log bench for our ease before the campfire. According to the topofil, Madzou reports, our day's progress has been 33,420 feet. Not a long walk, but a full one. After dark, as Fay and Madzou and I sit eating popcorn, there comes a weird, violent, whooshing noise that rises mystifyingly toward crescendo, and then crests — as, whoa, an elephant charges through camp, like a freight train with tusks. As the Pygmies dive for safety, sparks explode from the campfire as though someone had dropped in a Roman candle. Then, as quickly, the elephant is gone. Anybody hurt? No. Dinner is served and the pachyderm in the kitchen is forgotten, just a minor distraction at the end of a typical day on the Megatransect.

Fay spent the late 1980s at a site in the Central African Republic, doing fieldwork on lowland gorillas for his doctorate. One of the methods he used was the line-transect survey, which involved cutting straight trails through his study area, a multiple array of them in parallel, and then walking the trails repeatedly to locate and count gorilla nests. Line transects are a familiar tool among field biologists, most often used to estimate total abundance of some species within a large area by statistical derivation from a small, relatively convenient sample. Another of Fay's field methods, which proved more congenial to his disposition, was what he labeled the "group follow." With a trusted Pygmy tracker, he would follow one group of gorillas discreetly but persistently for all of one day or several, holding back at distance enough (several hundred yards) to leave them unaware of his presence. Such fastidious tracking allowed him to learn what they had been eating, how many nests they had built, and how to make deductions about group size, ages, gender, while minimizing the chance that he'd spook these very shy primates. Toward the end of his study he and his Pygmy mentor followed one group from dawn to dark for twelve days, resting and eating and walking in a shadowy rhythm with the gorillas. From a reading of his eventual dissertation, it seems the twelve-day follow was a high point in his academic fieldwork.

He returned to grad school in St. Louis meaning to write that dissertation, but after a few weeks he shoved it aside (not to be finished until eight years later) and flew back to Africa, seizing the irresistible distraction of more fieldwork. His new assignment was to do a survey of forest elephants in northern Congo. Fay was to focus the survey on three remote, difficult ecosystems: an area near the Gabonese border known as Odzala, a vast swampland to the east known as Likouala aux Herbes, and, farther north, a zone of trackless forest between the Nouabalé and Ndoki Rivers. Teaming up with an adventuresome Congolese biologist named Marcellin Agnagna, Fay set himself the delectable (to him) task of traversing all three areas on foot. A year later he returned for a second survey trek in the Nouabalé-Ndoki area, where he had found such a wonderland of unspoiled forest that it would eventually, after much determined but deft politicking by Fay and others, become one of Congo's most treasured national parks.

By 1991 Fay himself was director of this Nouabalé-Ndoki National Park project, on a management contract between the Republic of the Congo government and the Wildlife Conservation Society. Although his administrative duties had grown heavy and his political reach had lengthened, he still slid out for a two- or three-week reconnaissance hike whenever the necessity or the excuse arose. And soon after that he began to brainstorm about applying his leg-power approach on a whole different scale.

On the afternoon of Day 5 we enter Nouabalé-Ndoki National Park, crossing the Ndoki River in dugout canoes borrowed from the park's little boat landing, then paddling up a deep black-water channel through meadows of swaying *Leersia* grass, and continuing onward by foot. We spend the last hour before sunset walking through a rainstorm so heavy it fills the trail with a cement-colored flood. On Day 7 we skirt the perimeter of Mbeli Bai, a large clearing much frequented by elephants and gorillas. His first glimpse of this bai back in 1990, Fay tells me, was an ugly experience: He found six elephant carcasses, some with their tusks already hacked out, others left to rot until the extraction would be easier. The park hadn't yet been decreed, and poaching was rampant. In recent years the situation has much improved.

The park has also brought protection to giant trees of the species most valued for timber, such as *Entandrophragma cylindricum,* informally known as sapele, one of the premium African hardwoods.

Pointing to a big sapele, he says, "There's something you wouldn't see on the other side of the river" — that is, west of the park, where selective logging has already combed away the most formidable trees. Later he notes a mighty specimen of *Pericopsis elata,* far more valuable even than sapele. A log of *P. elata* that size is like standing gold, Fay says, worth serious money coming out of the sawmill. Spotting another, he changes his metaphor: "If sapele is the bread and butter around here, *Pericopsis* is the caviar."

We linger through mid-afternoon with a group of unwary chimpanzees that have gathered around us at close range — an eerie experience, given that most chimps throughout central Africa are terrified of humans, who commonly hunt them as food. These animals perch brashly in trees just above us, hooting and gabbling excitedly, sending down pungent but unmalicious showers of urine, scratching, cooing, ogling us with intense curiosity. One female holds an infant with an amber face and huge, backlit orange ears, neither mother nor baby showing any fear. A young chimp researcher named Dave Morgan, who has joined us for this leg of the hike, counts eleven individuals, including one with a distinctively notched left earlobe.

It's a mesmerizing encounter, for us and them, but after two hours with the chimps we push on, then find ourselves running out of daylight long before we've reached a suitable campsite. None of us wants a night without water. We grope forward in the dark, wearing headlamps now, cutting and twisting through kaka, finally stumbling into a sumpy, uneven area beside a muddy trickle, and Fay declares that this will do. Early next morning we hear chimps again, calling near camp. With Morgan's help we realize that it's probably the group from yesterday, having tracked us and bedded nearby. Camp-following chimps? The sense of weird and unearthly comity only increases when, on Day 8, we cross into an area known as the Goualougo Triangle.

At 4:15 that morning I'm awake in my tent, preparing for the day's walk by duct-taping over the sores and raw spots on my toes, ankles, and heels. To travel the way Mike Fay travels is hard on the feet, even hard on *his* feet, not because of the distance he walks but because of where and how. After a week of crossing swamps and stream channels behind him, I'm long since converted to Fay's notion of the optimal trail outfit — river sandals, shorts, one T-shirt that can be rinsed and dried. But the problem of foot care remains, partly because of the unavoidable cuts, stubs, and slashes inflicted by those thorny *H. danckelmaniana* stems and other hazards, partly because the sandy mud of Congolese swamps has an effect like sandpaper socks, chafing the skin away wherever a sandal strap binds against the foot. So I've adopted the practice of painting my feet with iodine every morning and night and (at the suggestion of another tough Congo trekker, a colleague of Fay's named Steve Blake) using duct tape to cover the old sores and protect against new ones. The stuff holds amazingly well through a day of swamp slogging, and although peeling off the first batch isn't fun, removal becomes easier on later evenings when there's no more hair on your feet. Since I've got a small roll of supple green tape, as well as a larger roll of the traditional (but stiffer, less comfortable) silver, I even find myself patterning the colors — green crosses over the tops of the feet, green on the heels, silver on the toes: a fashion statement. If my supplies of iodine and tape can be stretched for another week and my mental balance doesn't tip much further, I'll be fine.

At 4:30 A.M. I hear Dave Morgan, in the tent beside mine, beginning to duct-tape his feet.

Over breakfast Fay himself asks to borrow my tape for a few patches on his toes and heels. I give him the silver, selfishly hoarding the green. Then again we walk.

Demarcated by the Goualougo River on one side, the Ndoki River on another, the Goualougo Triangle is a wedge-shaped area extending southward from the southern boundary of Nouabalé-Ndoki National Park. Which means it's ecologically continuous with the park but not part of it statutorily, and it is isolated from the wider world by the two rivers.

The triangle embraces roughly 140 square miles of primary forest and swamp, including much excellent chimpanzee habitat, a warren of elephant trails, and an untold number of big sapele trees, all encompassed within a logging concession held by a company called Congolaise Industrielle des Bois (CIB), the largest surviving timber enterprise in northeastern Congo. With two sawmills, a shipyard, a community hospital, and logging crews in the forest,

CIB employs about 1,200 people, mostly in the towns of Kabo and Pokola, along the Sangha River. Although the company has shown willingness to collaborate with WCS in managing a peripheral zone south of the park, especially in restricting the commercial trade in bush meat, concern now seems to be gathering around the issue of the Goualougo Triangle. Mike Fay originally hoped to see that wedge of precious landscape included in the park, but when the boundaries were drawn in 1993, the Goualougo was lined out.

Logging operations were scheduled to begin in the Goualougo in 1998, but CIB agreed to postpone cutting and conduct an on-the-ground assessment of the timber resource and the costs of extracting it. That assessment — a *prospection*, in the jargon of Francophone forestry — will put a price tag on the triangle. Meanwhile the company, in a spirit that mixes cooperation with hardheaded bargaining, has welcomed WCS to participate in that prospection, focusing on assessment of the area's biological value. Weeks after returning from the Congo, I hear the CIB position put by the company's president, Hinrich Stoll. "You cannot just say, 'Forget about it; it is completely protected,'" he tells me by phone from his office in Bremen, Germany. "We all want to know how much it is worth." Once its worth has been gauged, both in economic and in biological terms, also in social ones, then perhaps the international community of conservationists and donors will see fit to compensate his company — yes, and the working people of Pokola and Kabo, Dr. Stoll stresses — for what it's being asked to give up.

But that talk of compensation, of balancing value against value, of ransoming some of the world's last ingenuous chimpanzees, comes later. As I stroll through the Goualougo with Fay, he turns the day into a walking seminar in forest botany, instructing me or quizzing Madzou and Morgan on the identity of this tree or that. Here's an *Entandrophragma utile,* slightly more valuable but far less common than its close relative *Entandrophragma cylindricum.* Its fruits resemble blackish yams festooned with wiry little roots, not to be confused with the banana-shaped fruit of another *Entandrophragma* species, *candollei.* And here's still another, *Entandrophragma angolense.* What about that tree there — what is it, Morgan? he demands. Um, an *Entandrophragma?* Wrong, Fay says, that one's *Gambeya lacourtiana.* Of course to me these are all just huge hulking boles, thirty feet around, rising to crowns in the canopy so high that I can't even see the shapes of their leaves. Morgan and Madzou are earnest students. Fay is a stern but effective teacher, sardonic one moment, lucid and helpful the next, drawing tirelessly on his own encyclopedic knowledge and his love for the living architecture of the forest. Now he directs Morgan's attention to the fine, fissured, unflaky bark of *Gambeya lacourtiana,* which is not to be confused with the more subtly fissured bark of *Combretodendron macrocarpum,* which is not to be confused with . . . a pile of lumber awaiting shipment from Kabo.

The good news from Day 8 is that Fay finds no *P. elata,* no standing gold, no caviar, at least along this line of march in the Goualougo Triangle. The bad news is that there's an abundance of the various *Entandrophragma* species, which are CIB's bread and butter. By the time the prospection team arrives to confirm or modify those impressions, Fay himself will be somewhere else, continuing his own singular sort of prospection at his own pace and scale.

From the triangle we make our way upstream along the Goualougo River, crossing back into the park. On the evening of Day 11 we're settled near an idyllic little bathing hole, a knee-deep pool with a sand bottom. Peeling away my duct-tape socks, after a gentle soak underwater, I feel exquisite relief. I wash my feet carefully, the rest of my body quickly, and then, given the luxury of deep, clear water, my hair. I rinse my shorts and T-shirt, wring them, put them back on. It's been a good day, enlivened by another two-hour encounter with a group of fearless chimps. For dinner there'll be a pasty concoction known as *fufu,* made from manioc flour and topped with some sort of sauce, plus maybe a handful of dried apricots for dessert. Then a night's blissful sleep on the ground; then fresh duct tape; then another day's walk. Having fallen into his rhythm, I've begun to see why Mike Fay loves this perverse, unrelenting forest so dearly.

Seated beside the campfire, Fay puts Neosporin antiseptic on his ragged toes. Several foot worms have burrowed in there and died, mortally disappointed that he wasn't an elephant. The ointment, as he smears it around, mixes with stray splatters of mud to make an unguent gray glaze. No, he affirms, there's no escaping foot hassles out here. You've just got to keep up the maintenance and try to avoid infection. When necessary, you stop walking for a few days. Lay up, rest. Let them heal. Wait it out. So he says. I can scarcely imagine what Fay's feet might have to look like before he resigns himself to that.

At the end of Day 13 we make camp on a thickly forested bench above the headwaters of the Goualougo, which up here is just a step-across stream. Our distance traversed since morning, as measured by the Fieldranger, is 42,691 feet. Our position is 2°26.297′ north by 16°36.809′ east, which means little to me but much to the great continuum of data. This particular day, alas and hoorah, has been my final one of walking with Fay, at least for now. (The plan is that I'll return later to share other legs of the hike.) Tomorrow I'll point myself toward civilization, retracing our trail of string and machete cuts to the Sangha River. Morgan and three of the Pygmies will accompany me.

And Fay? He'll continue northeastward and meet Nick Nichols, then loop down again through Nouabalé-Ndoki National Park before heading out across the CIB logging concessions and the other variously tracked and untracked forests of central Africa. The Megatransect has only begun: 13 days gone, roughly 400 to go. Many field notes remain to be taken, many video- and audiotapes to be filled, much data to be entered in computers, many miles of topofil to unroll. Then will come the challenge of making it all matter — collation, analysis, politics. When he reaches the seacoast of Gabon, Fay has told me, he'll probably wish he could just turn around and start walking back.

DONOVAN WEBSTER

Inside the Volcano

FROM *National Geographic*

THE VOLCANO'S SUMMIT is a dead zone, a cindered plain swirling with poisonous chlorine and sulfur gases, its air further thickened by nonstop siftings of new volcanic ash. No life can survive this environment for long. On the ash plain's edge, always threatening to make the island an aboveground hell, sit two active vents, Marum and Benbow, constantly shaking the earth and spewing globs of molten rock into the air. Yet across the black soil of the plain come all nine of us, a team of explorers, photographers, a film crew, a volcanologist, and me. We have hacked through dense jungles on this island called Ambrym, one of some eighty islands making up the South Pacific nation of Vanuatu, and entered this inhospitable landscape to camp and explore for two weeks. We've tightroped up miles of eroded, inches-wide ridgeline — with deep canyons plummeting hundreds of feet on either side — to totter at the lip of the volcanic pit of Benbow. The pit's malevolent red eye — obscured by gases and a balcony ledge of new volcanic rock — sits just a few hundred feet below.

"Okay, your turn," Chris Heinlein shouts above the volcano's roar.

A sinewy and friendly German engineer, Heinlein hands me the expedition's climbing rope, which leads down, inside the volcano. Clipping the rope into a rappelling device on my belt — which helps control my descent — I step into the air above the pit.

A dozen feet of rope slips between my gloved fingers. I lower myself into the volcano. Acidic gas bites my nose and eyes. The sulfur dioxide is mixing with the day's spitting drizzle, creating a sulfuric-acid rain so strong it will eat the metal frames of my eyeglasses within days, turning them to crumbly rust. The breathing of Benbow's pit is deafening, like up-close jet engines mixed with a cosmic belch. Each new breath from the volcano heaves the air so violently my ears pop in the changing pressure — then the temperature momentarily soars. Somewhere not too far below, red-hot, pumpkin-size globs of ejected lava are flying through the air.

I let more rope slip. With each slide deeper inside, I can only wonder: Why would anyone *do* this? And what drives the guy on the rope below me — the German photographer and longtime volcano obsessive Carsten Peter — to do it again and again?

We have come to see Ambrym's volcano close up and to witness the lava lakes in these paired pits, which fulminate constantly but rarely erupt. Yet suspended hundreds of feet above lava up to 2,200 degrees Fahrenheit that reaches toward the center of the Earth, I'm also discovering there's more. It is stupefyingly beautiful. The enormous noise. The deep, orangy red light from spattering lava. And those dark and brittle strands called Pele's hair: Filaments of lava that follow large blobs out of the pit, they cool quickly in the updraft and create six-inch-long, glassy threads that drift on the wind. It is like nowhere else on Earth.

Our first night on Ambrym we make camp in a beachside town called Port-Vato at the base of the 4,167-foot-high volcano. Shortly after sunrise the next morning, at the start of a demanding hike up the side of the volcano — walking a dry riverbed through thick jungle — I try to extract Peter's reasons for coming. As we crunch along the floor of black volcanic cinders, scrambling over shiny cliffs of cold lava that become waterfalls in the rainy season, Peter, forty-one, is grinning with excitement. Overhead dark silhouettes of large bats called flying foxes crease the morning sky like pterodactyls.

"I was fifteen years old and on vacation in Italy with my parents. They took me to see Mount Etna," he says. "As soon as I saw it, I was drawn to the crater's edge. I was *fascinated.* My parents went back to the tour bus. They honked the horn for me to come — but I couldn't leave. I edged closer, seeing the smoke inside, imagining the boiling magma below. At that moment I became infected."

Since then Peter has traveled the world examining volcanoes. His trips have taken him to Iceland, Ethiopia, Indonesia, Hawaii, and beyond. "And of course," he says, "I have been back to Etna, my home volcano, many times."

Using single-rope descending and climbing techniques developed by cave explorers and adapted for volcanoes, Peter has been dropping into volcanoes now for nearly a decade. "The size and power of a volcano is like nothing else on Earth," he says. "You think you understand the Earth and its geology, but once you look down into a volcanic crater and see what's there, well, you realize you will never completely understand. It is that powerful. That big." He grins. "You'll find out what I mean, I think."

After a five-hour walk uphill I get my first glimpse of that power as the expedition emerges from the steep, heavily vegetated sides of the volcano's cone and onto the caldera. In the course of a few hundred yards the trail flattens out, and the palm trees and eight-foot-tall cane grasses that lushly lined the trail behind us become gnarled and dead, their life force snuffed by a world of swirling gas clouds and acid rain.

This is Ambrym's ash plain. Seven miles across, it's a severely eroded ash-and-lava cap hundreds of feet thick. Across the plain Benbow and Marum jut almost a thousand feet into the sky.

To protect ourselves from the harsh environment, our team quickly establishes a base camp near the caldera's edge. Shielded behind a low bluff separating the caldera from the jungle, the camp stretches through a grove of palms and tree-size ferns, the black soil dotted with purple orchids bobbing on long green stalks. For the remainder of this first afternoon we set up tents and create acid-rain-tight storage areas. The camp is a paradise perched on the edge of disaster. As night falls, we eat chicken soup fortified with cellophane noodles and plan tomorrow's exploration, the volcano rumbling regularly in the background as we talk.

After dinner we follow Carsten Peter to the edge of the ash plain and watch the vents light the gas clouds, wreathing each peak in ghostly red glows. "Look there," Peter says, pointing to a third red cauldron halfway up Marum's side. "That must be Niri Taten. Tomorrow we'll start there."

All night long the rumbling keeps awakening us. Just a few miles away lava boils and the Earth roars while each of us — lying quietly in a flimsy tent — anxiously dreams of those swirling red clouds. Tomorrow night at this time, I resolve as I drift off to sleep once again, one thing is certain: It will have been a day like none I've ever had.

In the morning, shortly after a sunrise breakfast, we strike out toward Niri Taten, several miles uphill. As we follow dry and eroded riverbeds toward the volcanic cones, a gentle rain falls.

"What does Niri Taten mean?" I ask our local guide, Jimmy.

"Niri Taten is a small pig," he replies. "A small mad pig. A crazy pig. A small pig that causes trouble to men."

Haraldur Sigurdsson, one of the world's premier volcanologists, walks alongside me in the dry riverbed, examining sheer cliff faces. He points out strata of tephra, a mixture of volcanic material. By examining these layers, volcanologists can tell a volcano's level of activity. Larger and coarser tephra far from a volcanic pit means a more powerful volcano, since heavier matter is thrown farther as more explosive energy is supplied.

It's true. The closer we hike to the craters, the more the character of the riverbed beneath our feet changes from silty black grit to charcoal-size stones — not unlike old-time furnace clinkers. "Each volcano has its own chemical fingerprint," Sigurdsson says. "Each volcano's mineral and elemental content is different because of the nature of the volcano itself: its rock and the shape of its vent. It helps volcanologists a lot in their study.

"Like the Tambora eruption of 1815 in Indonesia," Sigurdsson says. "We've found Tambora ash by its particular chemical signature almost everywhere on Earth. One of that magnitude happens about every thousand years." The Tambora explosion is said to have given off so much ash and sulfur dioxide — both of which blocked and reflected sunlight — that 1816 was a "year without a summer" across much of the world. There was crop-killing frost throughout the summer in New England. In northern Europe harvests were a disaster.

Suddenly, from two miles upwind behind us, Benbow gives a huge belch. We turn to look back. "Uh-oh" Sigurdsson says. "Ashfall on the way." Instead of the usual bluish white clouds of steam and gas, the plume issuing from the cone is heavy and black, trailing earthward in a dark curtain. Slowly it drifts our way on the wind. Five minutes later the ashfall finds us, covering our rucksacks, clothing, faces, boots, and ponchos with a sandy grit the color of wet cocoa mix.

Under the ashfall we climb Marum, pressing forward through the dead volcanic soil for another hour. Each step takes us closer to

Niri Taten, a crater that tunnels straight down into the basaltic rock like a massive, steaming worm burrow 200 yards across. As we approach, a rising wind and thick clouds of chlorine gas force us to pause and pull on safety helmets and industrial-style gas masks that cover our noses and mouths. Without them, between the flying bits of stone and grit carried on the fifty-mile-an-hour winds and the thick clouds of gas roaring upward from the vent, time spent near the pit's lip would be painfully dangerous if not impossible.

Even with these protections the howling wind and gas often force us to shut our eyes and suspend breathing until the heaviest gas clouds pass. We lean against the high winds, brace at the crater's edge, and look inside.

Five hundred feet below, the vent's opening is obscured by rocky ledges. But if we can't see the lava itself, there is a consolation. Every inch of rocky surface inside the vent's cone is painted with color. Sunshine yellow sulfur coats some of the crater's sheer rock faces. Iron washes other sections of rock with flaming orange. Pastel green deposits of manganese glaze rock nearest the vent, like a carpet of immortal moss. Other patches of stone have been bleached white by chlorine and fluorine gases pouring from the vent.

Besides the wind and dangerous concentration of gases, the edges of Niri Taten are too crumbly to allow safe descent. Anyone climbing down a rope inside the crater could be dislodging loose boulders, some the size of cars, that could crash on anyone below. Carsten Peter pulls out his camera and long lens — whose coating immediately becomes corroded in the noxious air. The howling gusts twice knock expedition members to the ground.

After an hour it's decided that we should examine the Marum crater itself. "We can get two volcanoes in one day!" Carsten Peter says with glee. Our helmeted heads tucked down, we continue breathing scuba-diver slow into our masks for maximum benefit, and we push on.

The walk to Marum's opening isn't far, but what it lacks in distance it makes up for in danger. No matter which route you choose, you have to traverse the mountain's steep slopes, many of which are gouged with deep, unclimbable erosion gullies. We decide to cross where the gullies are smallest: along Niri Taten's knife-edged lip, within a foot of a sheer drop into the crater.

We step gingerly where the slope looks most reliable, but our footing remains dangerously slick. The slope's top layer is crumbly tephra, sometimes as big as charcoal briquettes. Making things more difficult, we've moved downwind of Niri Taten. All around us clouds of sulfur dioxide, chlorine, and fluorine gases swirl so thick they sometimes obscure our vision and force us to stop and bury our gas-masked faces inside our arms for extra protection.

It's a slog. Minutes stretch into an hour. Every step could be our last. Finally we reach the summit of the crater's edge and begin down its other side. Protected by the lip behind us, the environment changes. Sunshine blankets the tilting black ash, and the cold gales calm into balmy breezes.

Two expedition members, Franck Tessier and Irène Margaritis, hustle downslope with me toward Marum. As we approach its lip, the thirty-nine-year-old Tessier — a genial and easygoing French biologist with impressive rope and rock-climbing skills honed by years of adventures like this one — rips off his gas mask and begins to hoot with pleasure.

I know why. Ahead of us Marum's volcanic pit stretches as open and clear as a visionary's painting. In the pit, three step-down ledges — each deeper and wider than the one uphill of it — are marbled with layers of black ash and pale, bleached basaltic andesite. The layers of lava inside the vent form as a crust over a cooling lava lake that gets blown out like a massive champagne cork when volcanic activity resumes. Small wall vents called fumaroles — created where heated groundwater and escaping volcanic gas reach the surface — let off steady plumes of steam. Inside Marum's crater it looks as if the world is being born.

And there, in the bottom of the third and largest pit — some 1,200 feet below — sits the lava lake. Its fury pushes lava through three skylight holes in a roof that partly covers the lake like a canopy. Bright orange and red spatters fly unpredictably from the circular opening of the largest skylight, a hole perhaps fifty yards across.

Lava is three times as dense as water. Despite its up to 2,200-degrees-Fahrenheit heat, lava moves, burbles, and flies through the air with the consistency of syrup. Every few minutes huge molten blobs seem to soar in slow motion. A second or two later a noise from beneath the earth — a rumbling *booooom* — fills the pit and rolls across the sculpted ash plain beyond. It's mesmerizing: lava sloshing back and forth, bubbles emerging and popping like a thick stew. As we survey Marum's lip and crater, I can't take my eyes off the lava. Suddenly I understand Peter's obsession. As evening cloaks the pit's deepest recesses in shade, the lava lake and explosive bubbles glow more seductively. The spatterings glisten like enormous, otherworldly fireworks as they sail through the shadowed air.

Dangling inside Benbow's crater the following afternoon, I have time to reflect. This morning we followed the narrow ridge to Benbow's pit — which was firm enough to climb down. We fixed our rope, ate lunch in a spitting acid rain, and began our descent into the volcano.

Now, on the rope below me, Carsten Peter works his way deeper inside the crater. I let more rope slide through my hands, easing myself deeper as well.

With each drop the air shakes more violently; the clouds of poison gas grow thicker. Waves of pressurized air rumble past me.

Grasping the rope tightly, I halt my descent at the edge of an overhung cliff and stare deeper inside. The lava lake waits below, ejecting orange bombs and smaller drops. Then, in a heartbeat, a wall of thick clouds blows between me and the pit, enveloping everything around me in a world of gray. In the shuddering air and disorienting noise, gravity, direction, and time seem to fade away. There is only the volcano, its existence a direct result of two tectonic plates colliding below me. Benbow roars again. The earth shakes.

In this moment I know I've gotten close enough to the fire at the center of the Earth. At that same second the clouds part and Benbow reappears. Fumaroles smoke, and steam swirls from the pit's walls. The Technicolor wash swarms around me like a kaleidoscope. Below, Carsten Peter hits the end of the rope just above Benbow's explosive vent. He pulls a camera from his bag and lifts it to his eye.

Torrential rains will frustrate another attempt to explore Benbow. Then dissension breaks out among some of the expedition's porters who helped carry gear up the volcano's steep cone, and it becomes clear that the team will have to leave Ambrym as soon as possible. In a last-ditch, eighteen-hour marathon, team members drop 1,200 feet into Marum and photograph its lava lake nonstop. They emerge from the crater and find a fractious camp. Jimmy cannot persuade the disgruntled porters to bend, and the tension escalates. With a satchel full of photographs, Carsten Peter finally agrees to abandon the volcano — even as he vows to return.

Contributors' Notes

Other Notable Science and Nature Writing of 2000

Contributors' Notes

David Berlinski earned his Ph.D. in analytic logic and philosophy from Princeton University. He is the author of several screenplays, three novels, and four works of nonfiction, including *Newton's Gift, A Tour of the Calculus,* and *The Advent of the Algorithm,* from which this article was adapted. He lives in Paris, in an apartment overlooking the Seine, where he is at work on a biography of Einstein.

Mark Cherrington is the editor of *Earthwatch* magazine, the author of *Degradation of the Land,* and a contributor to *Masterworks of Man and Nature.* His work has taken him on expeditions to track rhinos, capture Komodo dragons, and dodge polar bears, and has appeared in *Discover, Audubon, Time, Vogue, Wildlife Conservation,* and the *Boston Globe,* among other publications.

A former editor of *The Sciences,* **Edwin Dobb** currently is a contributing editor of *Harper's Magazine.* He is the coauthor with Jack Horner of *Dinosaur Lives: Unearthing an Evolutionary Saga* and teaches magazine writing at the UC Berkeley Graduate School of Journalism.

Gregg Easterbrook is senior editor of the *New Republic,* contributing editor of *The Atlantic Monthly,* and a visiting fellow at the Brookings Institute. His most recent book is *Beside Still Waters.* He lives in Bethesda, Maryland.

Malcolm Gladwell is the author of *The Tipping Point.* From 1987 to 1996, he was a reporter for the *Washington Post,* first as a science writer and then as New York City bureau chief. Since 1996, he has been a staff writer for *The New Yorker.*

Jane Goodall received her Ph.D. from Cambridge University. Her research on chimpanzees has been described by biologist Stephen Jay Gould as "one of the Western world's great scientific achievements." Her books include *In the Shadow of Man, Reason for Hope,* and, most recently, *Beyond Innocence.* She resides in Tanzania.

Jerome Groopman, M.D., is the Recanati Professor of Immunology at Harvard Medical School, chief of experimental medicine at Beth Israel Deaconess Medical Center, and one of the world's leading researchers in cancer and AIDS. He and his work have been featured in the *New York Times,* the *Wall Street Journal,* the *Boston Globe,* and *The New Yorker,* as well as in numerous scientific journals. He lives with his family in Massachusetts.

Stephen S. Hall is a contributing writer for the *New York Times Magazine* and writes frequently about science and medicine for *Technology Review* and other magazines. He is the author of three books: *Invisible Frontiers, Mapping the Next Millennium,* and *A Commotion in the Blood.* He lives in Brooklyn, New York, with his wife and two children.

Bernd Heinrich is the author of twelve books, most of them dealing with his biological research and field observations. In high school he "majored" in cross-country running, but he went on to earn a Ph.D. in zoology from UCLA. This essay was adapted from his most recent book, *Racing the Antelope.*

Edward Hoagland lives in Bennington, Vermont. He began his first novel, *Cat Man,* fifty years ago, and it won a Houghton Mifflin literary fellowship. His seventeenth book, a memoir entitled *Compass Points,* was published this year.

Bill Joy, chief scientist at Sun Microsystems, which he cofounded in 1982, was cochair of a presidential commission on the future of Information Technology Research under President Clinton and has received many computer industry awards recognizing his work. He is currently working on a book-length treatment of the issues raised by his essay.

Ted Kerasote's nature writing has appeared in more than fifty periodicals, including *Audubon, National Geographic Traveler, Outside,* and *Sports Afield,* where his Environment column follows the many issues of wildlife and wildlands conservation. The author of three books — *Navigations, Bloodties,* and *Heart of Home* — he is also the editor of the Pew Wilderness Center's annual, *Return of the Wild.*

Barbara Kingsolver was trained as a biologist and has published nine books, including the award-winning novels *Prodigal Summer* and *The Poisonwood Bible.* **Steven Hopp** is an ornithologist at the University of Arizona and a research associate at the Arizona-Sonora Desert Museum. His primary research is on vocal communication in vireos and other birds. With their two daughters, they divide their time between Arizona and southern Appalachia.

Verlyn Klinkenborg is the author of *Making Hay* and *The Last Fine Time.* He is a member of the New York Times Editorial Board and lives with his wife in rural New York State.

Jon R. Luoma is a contributing editor at *Audubon* magazine, and his work has appeared in many other national magazines including *Life, National Geographic, Wildlife Conservation, Discover,* the *New York Times Magazine,* and *GQ,* as well as the Tuesday Science Times section of the *New York Times.* His most recent book is *The Hidden Forest.*

Cynthia Mills is a practicing dog and cat doctor and an occasional researcher. She implemented a nationwide clinical trial treating cancer in companion animals, and, more recently, spent a summer tramping through forests in Fairfax County, Virginia, catching raccoons for an oral rabies vaccination project. She writes about science in her spare time.

Oliver Morton writes about science and technology and their effects. Formerly science and technology editor at the *Economist,* he is now a contributing editor at *Wired* magazine; he has also written for *The New Yorker, Nature, Science, Prospect,* and the *Hollywood Reporter* (but only once . . .). His first book, *Mapping Mars,* will be published in spring 2002.

Val Plumwood is an ARC Fellow at the University of Sydney and author of *Feminism and the Mastery of Nature,* which one reviewer described as "shaking philosophy to its foundations." She lives at the edge of the coastal escarpment of the southern tablelands of New South Wales, learning from and being nurtured by the mysteries and marvels of the surrounding forest.

Sandra Postel is author of *Pillar of Sand* and *Last Oasis,* which was the basis for a PBS documentary that aired in 1997. She directs the Global Water Policy Project in Amherst, Massachusetts, where her research focuses on international water issues and strategies. She is also a senior fellow with Worldwatch Institute, a visiting senior lecturer in environmental studies at Mount Holyoke College, and a Pew Fellow in Conservation and the Environment.

Richard Preston is the author of *The Cobra Event, The Hot Zone, American Steel,* and *First Light.* He is a contributor to *The New Yorker* and has won numerous awards, including the McDermott Award in the Arts from MIT, the American Institute of Physics Award in science writing, and the Overseas Press Club of America Whitman Basso Award for best reporting in any medium on environmental issues.

David Quammen is the author of *The Song of the Dodo* and eight other books of nonfiction and fiction. For fifteen years he was a columnist for *Outside* magazine, and twice received the National Magazine Award for his science essays and other work there. In 1996 he received an Academy Award in literature from the American Academy of Arts and Letters. His most recent book is *The Boilerplate Rhino,* a collection of essays.

Donovan Webster is the author of the award-winning book *Aftermath: The Remnants of War.* A regular contributor to *National Geographic, Smithsonian,* and *Men's Journal,* he lives with his wife and children in central Virginia. His book *The Burma Road,* a history of World War II's China-Burma-India Theater of Operations as viewed from those locations today, will be published next year.

Other Notable Science and Nature Writing of 2000

SELECTED BY BURKHARD BILGER

NATALIE ANGIER
Biological Bull. *Ms.*, June/July.

KIM BARNES
The Ashes of August. *Georgia Review*, Summer.
RICK BASS
The Winter's Tale. *The Atlantic Monthly*, January.
LISA BELKIN
The Making of an 8-Year-Old Woman. *The New York Times Magazine*, December 24.
JEREMY BERNSTEIN
Creators of the Bomb. *The New York Review of Books*, May 11.
DAVID BERREBY
Race Counts. *The Sciences*, September/October.
WENDELL BERRY
Life Is a Miracle. *Orion*, Spring.
SUSAN BLACKMORE
The Power of Memes. *Scientific American*, October.
DAVID BODANIS
The Fires of the Sun. *Wilson Quarterly*, Summer.
DEBBIE BOOKCHIN AND JIM SCHUMACHER
The Virus and the Vaccine. *The Atlantic Monthly*, February.
GUY C. BROWN
Speed Limits. *The Sciences*, September/October.
JEROME BRUNER
Tot Thought. *The New York Review of Books*, March 9.
ALAN BURDICK
Name That Star! *Discover*, February.

CLAUDE R. CANIZARES
X-Ray Visionaries. *The Sciences,* May/June.
ALSTON CHASE
Harvard and the Making of the Unabomber. *The Atlantic Monthly,* June.
JON COHEN
The Hunt for the Origin of AIDS. *The Atlantic Monthly,* October.
SIMON COLE
The Myth of Fingerprints. *Lingua Franca,* November.
RICHARD CONNIFF
Jelly Bellies. *National Geographic,* June.
So Tiny, So Sweet . . . So Mean. *Smithsonian,* September.

KEITH DEVLIN
Snake Eyes in the Garden of Eden. *The Sciences,* July/August.
BILL DONAHUE
Pilgrim at Johnson Creek. *DoubleTake,* Spring.

HELEN EPSTEIN
Bonobos in Paradise. *Lingua Franca,* October.
CAROL EZZELL
Care for a Dying Continent. *Scientific American,* May.

LESLIE LEYLAND FIELDS
Harvester Island. *Orion,* Winter.
ROBERT FINCH
On the Killing Fields. *Georgia Review,* Spring.
MARIA FINN
Pigeon Mumblers. *Chicago Review,* no. 2.
MARSHALL JON FISHER
Bodysurfing: My Father's Board. *DoubleTake,* Summer.
TIM FOLGER
From Here to Eternity. *Discover,* December.
BRENDA FOWLER
The Iceman Melteth. *Lingua Franca,* May/June.
IAN FRAZIER
Tomorrow's Bird. *DoubleTake,* Fall.

PATRICIA GADSBY
Tourist in a Taste Lab. *Discover,* July.
MALCOLM GLADWELL
John Rock's Error. *The New Yorker,* March 13.
The New-Boy Network. *The New Yorker,* May 29.
ADAM GOODHEART
The Last Island of the Savages. *American Scholar,* Autumn.
MIKHAIL GORBACHEV
The Face of the Waters. *Civilization,* October/November.
DANIEL GOTTESMAN AND HOI-KWONG LO
From Quantum Cheating to Quantum Security. *Physics Today,* November.

Charles Graeber
How Much Is That Doggie in the Vitro? *Wired,* March.
Steve Grenard
Is Rattlesnake Venom Evolving? *Natural History,* July/August.
Gordon Grice
Where Leprosy Lurks. *Discover,* November.
Jerome Groopman
Second Opinion. *The New Yorker,* January 24.

Jeffrey A. Hammond
Science Boy. *Salmagundi,* Spring/Summer.
Marc D. Hauser
What Do Animals Think About Numbers? *American Scientist,* March/April.
Theodore P. Hill
Mathematical Devices for Getting a Fair Share. *American Scientist,* July/August.
Jack Hitt
The Second Sexual Revolution. *The New York Times Magazine,* February 20.
Edward Hoagland
Fire. *American Scholar,* Autumn.
Marguerite Holloway
The Killing Lakes. *Scientific American,* July.
Wolves at the Door. *Discover,* June.

George Johnson
The Jaguar and the Fox. *The Atlantic Monthly,* July.
Wendy Johnson
Wild Iris. *Tricycle,* Summer.

Jamie Lincoln Kitman
The Secret History of Lead. *The Nation,* March 20.
Dan Koeppel
The List. *Audubon,* September/October.
Elizabeth Kolbert
The River. *The New Yorker,* December 4.
Robert Kunzig
Gluons. *Discover,* July.

Jacques Leslie
Running Dry. *Harper's Magazine,* July.
Barry Lopez
Informed by Indifference. *Whole Earth,* Spring.

Dana Mackenzie
May the Best Man Lose. *Discover,* November.
Kristie Macrakis
The Case of Agent Gorbachev. *American Scientist,* November/December.
Scott L. Malcomson
The Color of Bones. *The New York Times Magazine,* April 2.

William H. McNeill
The Flu of Flus. *The New York Review of Books,* February 10.
John McPhee
A Selective Advantage. *The New Yorker,* September 11.
They're in the River. *The New Yorker,* April 10.
Oliver Morton
The Computable Cosmos of David Deutsch. *American Scholar,* Summer.

Gary Paul Nabhan
Coming Home to Eat. *Orion,* Summer.
Vladimir Nabokov
Father's Butterflies. *The Atlantic Monthly,* April.
Spencer Nadler
Brain-Cell Memories. *Harper's Magazine,* September.
Early Alzheimer's: A View from Within. *Missouri Review,* Winter.
Emily Nussbaum
A Question of Gender. *Discover,* January.
Martha C. Nussbaum
Brave Good World. *The New Republic,* December 4.

Mary Odden
The Sound of a Meadowlark. *Georgia Review,* Spring.
Danielle Ofri
M&M. *Missouri Review,* Winter.
Lawrence Osborne
A Stalinist Antibiotic Alternative. *The New York Times Magazine,* February 6.
Dennis Overbye
On the Eve of 2001, the Future Is Not Quite What It Used to Be. *The New York Times Magazine,* December 26.
Quantum Theory Tugged, and All of Physics Unraveled. *The New York Times,* December 12.

Jake Page
Making the Chips That Run the World. *Smithsonian,* January.
Robert L. Park
Welcome to Planet Earth. *The Sciences,* May/June.
Doug Peacock
The Voices of Bones. *Outside,* February.
Henry Petroski
Vanities of the Bonfire. *American Scientist,* November/December.

David Quammen
The Post-Communist Wolf. *Outside,* December.

Matt Ridley
Asthma, Environment and the Genome. *Natural History,* March.
Joseph Roman
Fishing for Evidence. *Audubon,* January/February.

Is the Right Whale Going Down? *Wildlife Conservation,* May/June.

JONATHAN ROSEN

Birding at the End of Nature. *The New York Times Magazine,* May 21.

ROBERT M. SAPOLSKY

Genetic Hyping. *The Sciences,* March/April.

KIRK SEMPLE

A Habitat Held Hostage. *Audubon,* November/December.

VANDANA SHIVA

A Worldview of Abundance. *Orion,* Summer.

EARL SHORRIS

The Last Word. *Harper's Magazine,* August.

CHARLES SIEBERT

Sentenced to Nature. *The New York Times Magazine,* December 17.

FLOYD SKLOOT

The Painstaking Historian. *Northwest Review,* vol. 38, no. 1.

MICHAEL SPECTER

The Pharmageddon Riddle. *The New Yorker,* April 10.

MARC K. STENGEL

The Diffusionists Have Landed. *The Atlantic Monthly,* January.

RICHARD STONE

Vostok. *Smithsonian,* July.

JOEL L. SWERDLOW

Unlocking the Green Pharmacy. *Wilson Quarterly,* Autumn.

MARGARET TALBOT

The Placebo Prescription. *The New York Times Magazine,* January 9.

GARY TAUBES

The Cell-Phone Scare. *MIT's Technology Review,* November/December.

MATTHEW TEAGUE

Out of This World. *Oxford American,* September/October.

JOHN TERBORGH

In the Company of Humans. *Natural History,* May.

RANDY THORNHILL AND CRAIG T. PALMER

Why Men Rape. *The Sciences,* January/February.

SALLIE TISDALE

Rescuing Orchids. *Audubon,* September/October.

MICHAEL S. TURNER

More Than Meets the Eye. *The Sciences,* November/December.

JOHN UPDIKE

A Layman's Scope. *Natural History,* February.

HANS CHRISTIAN VON BAEYER

The Lotus Effect. *The Sciences,* January/February.

ALLISON WALLACE

The Honeybee's Incredible, Edible Metaphor. *Orion,* Summer.

Chip Ward
Cowboys in Gas Masks. *Orion,* Summer.

T. H. Watkins
High Noon in Cattle Country. *Sierra,* March/April.

Richard White
Dead Certainties. *The New Republic,* January 24.

Jonathan Weiner
Curing the Incurable. *The New Yorker,* February 7.

Ted Williams
Living with Wolves. *Audubon,* November/December.

Richard Wolkomir
Charting the Terrain of Touch. *Smithsonian,* June.

window
2246615